D0303870

Accessibility and Utilization

Accessibility and Utilization

Geographical perspectives on health care delivery

Alun E. Joseph
Department of Geography
The University of Guelph
Ontario, Canada

David R. Phillips
Department of Geography
The University of Exeter
Exeter, England

HARPER & ROW, PUBLISHERS
NEW YORK

Cambridge
Hagerstown
Philadelphia
San Francisco

London
Mexico City
Sao Paolo
Sydney

LIVERPOOL INSTITUTE
OF HIGHER EDUCATION

THE MARKLAND LIBRARY

Accession No.

88014

Class No. 911.12
JOS

Catal. UM
4/10/84

Copyright © 1984 A.E. Joseph and D.R. Phillips
All rights reserved

First published 1984
Harper & Row Ltd
28 Tavistock Street
London WC2E 7PN

No part of this book may be reproduced in any manner
whatsoever without written permission except in the case of brief
quotations embodied in critical articles and reviews.

British Library Cataloguing in Publication Data
Joseph, A.E.
    Accessibility and utilization: geographical
    perspectives on health care delivery
    1. Medical care
    I. Title   II. Phillips, D.R.
    362.1   RA393

    ISBN 0-06-318276-9

Typeset by Burns & Smith, Derby
Printed and bound by Butler & Tanner Ltd. Frome and London

About the Authors

Alun Joseph is Associate Professor of Geography at the University of Guelph, Ontario. His research interests are in rural geography and the application of quantitative methods to the analysis of service provision and location. He has written numerous articles appearing in journals such as *Economic Geography,* the *Canadian Geographer* and *Social Science and Medicine.*

David Phillips is Lecturer in Geography at the University of Exeter. He has also held visiting appointments in the University of Hong Kong and the University of the Philippines. His research interests are in medical, urban and social geography. His publications include *Contempory Issues in the Geography of Health Care* (Geo Books, Norwich), as well as articles on health care delivery in a range of professional journals.

CONTENTS

Preface

Chapter One

Chapter Two

Chapter Three

Chapter Four

Physician organization, location and access to health care

Chapter Five

Measuring the potential physical accessibility of general practitioner services

Chapter Six

Utilization of health care facilities: revealed accessibility?

Chapter Seven

Chapter Eight

Preface

Service availability and utilization have become important aspects in the teaching of many urban, rural or social geography courses. Medical and health care services are frequently cited as examples of services to which consumer access is critical and the study of this general topic also forms a major focus of the well-established subject of medical geography. This book is intended as a text for use in undergraduate university and college courses in these subjects. It will also, we hope, form a useful text for wider courses in geography in which service location and utilization form a component. In addition, it has been written very much as a resource book to which non-specialist geographers, social scientists and professionals involved in the provision of health and other public services can refer. Therefore, we have taken pains to explain carefully spatial principles involved at various points in the book and sought to provide from our varied geographical backgrounds in institutions on different sides of the Atlantic as full and comprehensive a bibliography as possible. Whilst we have referred to the literature widely, we have also taken advantage of our own research experiences in this subject to provide illustrations of general propositions, which will account for the nature of examples cited.

This is a text in medical geography but its concern is not with the epidemiology and ecology of diseases but with the location and utilization of health services. It is, of course, a moot point whether these two aspects of medical geography can or should be separated, which is discussed further in Chapters 1 and 8. We regard Chapter 2 as an important factual and benchmark statement which analyses the nature and evolution of health services. It provides specific examples which have emerged under different political and economic systems. Indeed, the political economy approach to understanding the nature of health services as part of societies is crucial to modern social scientific analysis. Chapters 3, 4 and 5 analyse the nature of access to health services, and particularly the effects of developments in physician organization on this. Chapters 6 and 7 concentrate on the effects which service type and accessibility, amongst other factors, have on their use by the public, the latter chapter focusing on the special case of services for mentally ill persons. It will be evident, as discussed in Chapter 1, that we have not considered in great detail the types of optimizing solutions to service location which were at one time in vogue. Instead, we hope that we have concentrated more on the practical matters of health care utilization and planning which are often pragmatic and policy oriented.

The book was conceived and produced very much as a joint effort as each of us perused, added to and amended the work of the other. Individual statements and opinions may remain but, on the whole, we feel that the book as it stands represents a fair balance of our views of the problems and prospects of accessibility and utilization of health services. We are aware, too, that this usually implies consideration of 'ill-health' services, because preventive medicine and the role of numerous other factors in society, such as employment, housing, nutrition and education, are probably as important (if not more important) to good health as are health services themselves.

Our interest in the subject matter of the book has been long-standing. David Phillips' interest grew from doctoral research and was further developed by research in Britain and abroad into health services planning and utilization. He is particularly grateful to have been able to be a member of the Exeter and District Community

Health Council since 1980 which has given him a particular insight into the local problems of health care planning and the role of the consumers' voice in this. Alun Joseph started researching in the Peterborough area of Ontario into distance decay effects in mental health care. Further work in health care planning and utilization was sponsored by the Ontario Ministry of Agriculture and Food as part of a programme of research on rural living. This provided additional empirical experience of practical health care planning and utilization. Both of us have been fortunate to be able to teach aspects of medical geography in specialist courses or as components of wider social geography courses in our respective University departments.

We wish to acknowledge the permission provided by various publishers to reproduce copyright material and these are noted as sources under specific figures in the text. However, we owe an immense debt of gratitude to a number of individuals who helped in the production of the book. Mrs Jane Hayman has, on this occasion as on many others, read most carefully and assiduously our manuscript and has given most helpful grammatical guidance as well as sensible comment on textual matters. In Exeter University Geography Department drawing office, Terry Bacon worked with great skill and talent to produce what we think to be a very competent and clear set of diagrams. Andrew Teed undertook the production of final photographic prints. We are also most grateful for secretarial assistance in typing manuscripts in both Departments but, in the University of Guelph in particular, Mrs Becky Morrison and Ms Rosemary Sexsmith had the daunting task of putting our text onto a word processor and conducting numerous editorial refinements. Last but not least, we are most grateful for the tolerance shown to us by our families during the writing and completion of the finished book.

Alun E. Joseph   David R. Phillips
Department of Geography   Department of Geography
The University of Guelph   The University of Exeter
Ontario, Canada   Exeter, England

February 1984   February 1984

L. I. H. E.
THE MARKLAND LIBRARY
STAND PARK RD., LIVERPOOL, L16 9JD

1   Geographical Perspectives on Health Care

Introduction and structure of the book

Medical geography is one of the oldest branches of geographical study and has incorporated, in true environmentalist style, a consideration of both the physical and human factors which combine to influence man's bodily and mental health. This subject area also continues to be a fertile one in applied geography as developments in medical and social science have indicated new directions for research into both the spatial incidence of disease and the provision of caring facilities. Medical geography is a constantly evolving subdiscipline, one which is worthy of sustained research interest. It also provides the student of geography and the general reader with an opportunity to observe the practical utility of an academic discipline and, indeed, to relate personal knowledge of the real world to the picture portrayed in academic textbooks.

This book is concerned with geographical aspects of health care systems. Within this broad orientation, a major focus is on primary health care, that level of care provided in 'Western' countries by community nurses and general practitioners, supposedly readily accessible and available to their patients, as discussed in Chapter 3. Such a focus was selected in spite of the knowledge that this level of care does not take up the majority of health services expenditure. Indeed, in almost every health care system hospital services consume a disproportionate amount of expenditure in relation to the number of cases treated but, here, less attention is devoted to this level of care. This book very much accepts modern thinking that only the primary level can extend some sort of cost-effective care to the vast bulk of populations, especially in developing countries. The current emphasis on primary health care (World Health Organization 1978a) is, like calls for 'appropriate technology', only making an old and established concept look more appealing. The important thing is that, to be successful, primary health care must reach out to all those in need of it and, to be effective, must achieve 100 percent outreach (Fendall 1981). This presents major problems of accessibility and utilization.

The book accords with contemporary developments in geography as a synthesizing social science. A major focus for this role is furnished by the 'welfare approach', which considers people, their access to and use of essential services, and their quality of life (Smith 1974, 1977, 1979; Coates et al. 1977; Herbert and Smith 1979). Medical services are undoubtedly amongst the most important social facilities in terms of accessibility. They are mostly curative but also form part of wider preventive and educative networks which help to maintain health and avoid ill-health. Differential patterns of access to health care facilities provide an excellent example of the way in which the spatial allocation of resources in societies can serve to intensify general disparities in the quality of life (Knox 1982a).

The authors are aware that some might criticize the use of the term 'health care' in the title as it suggests a curative orientation, more readily identifiable as 'care in the event of illness'. Such critics would argue that 'ill-health care' is but one aspect of total health care, which should include a study of housing, nutrition, education, employment, politics, and many other topics, all of which underlie 'quality of life' and together promote or damage 'health'. To adopt a physician- or facility-oriented approach is therefore to achieve a limited perspective which, being only partial, will leave a distorted image in the mind of the reader. In addition, some critics such as Illich (1976) go further and suggest that many health problems are actually iatrogenic, or physician-created. These criticisms must be admitted, at least in part, but their total acceptance would imply the adoption of an overidealistic attitude at the moment. Medical facilities and personnel have to be distributed in accordance with the objectives of medical systems; modern medicine with some evil side-effects exists, and it will continue to be used and facilities attended by most intending patients. This book has to limit its focus to the effects which spatial and social factors have on this process even if, in doing so, other variables in the behavioural and institutional networks encompassing health care receive lesser attention.

Accessibility can be of two main types, physical and socio-economic. The former implies that a service and the means of reaching it are available (Moseley 1979), whilst socio-economic accessibility involves people's ability to pay for a service, whether they feel it is appropriate (introducing psychological and perceptual-evaluative facets), and whether they are permitted to use it (organizational and institutional restrictions on accessibility). Therefore, in Chapters 4 and 5, the focus is on physical availability, more general considerations of access to medical services having been introduced in Chapter 3. Chapter 5, in particular, examines two important measures of potential access to services, which involve the delimitation of regional availability of services and regional access to them.

There are, manifestly, social and spatial differences in access to medical services. The question of socio-economic access really involves the identification of variables which affect *utilization* of services. These could be factors which promote utilization by some people – such as age, sex, mobility, income, or knowledge – or which hinder use by others – perhaps lack of income or mobility. Since the social and demographic mix of areas of cities and countries varies greatly, this composition may well be reflected in different utilization patterns over space. It is a very complex question but differential availability and differential utilization of services have become important foci for social geographical research in general (Thomas 1976; Herbert and Thomas 1982; Knox 1982a); and medical geographical research in particular (Haynes and Bentham 1979; Phillips 1979a, 1981a; Health Research Group 1982; Whitelegg 1982; Eyles and Woods 1983).

It is sensible to consider access in terms of whether or not those who need care obtain it (Aday and Andersen 1974). 'The proof of access is use of service, not simply the presence of a facility. Access can, accordingly, be measured by the level of use in relation to "need"...' (Donabedian 1973, p.211). In this way, we argue that utilization can be taken to be *revealed* accessibility. Chapter 6 discusses the variables which geographers and other social scientists have found to influence utilization over a range of health care services (maintaining the primary care interest, of course), whilst Chapter 7 examines the special case of use of mental health care facilities. This is an old focus of geographical concern and, sadly, one which is growing in importance today because, in developed countries and increasingly so in developing countries, mental illness is becoming a major cause of contact with health services.

In Chapter 2, we take the position that the nature of the health care system which exists in any given nation is determined as much by the political economy of the country in question as by the level of economic development and the nature of medical technology available. A nation's health care system should be viewed as an important parameter which determines whether individuals are able to use facilities; it acts as a constraint in the access it allows (spatially and economically) and in the people to whom it permits use (whether there is universal or selective coverage). Therefore, in Chapter 2, we have attempted to cover the spectrum of 'health care delivery systems' (the jargon or shorthand term), selecting examples of countries with differing systems at different levels of economic development and different political leanings. The nature and evolution of systems, especially their financial bases and spatial distribution, are highlighted, with the overall aim of the chapter being to explain the influence these have on current availability patterns and on utilization by patients or potential patients. Our focus therefore is very much on the way in which different health care systems influence utilization in broad terms. This provides a foundation of empirical examples for succeeding theoretical sections of the book. In particular, we have taken pains to point out that modern 'Western' medical systems, which have to date formed the main topics of geographical interest, are by no means universal or always available. Therefore, the nature of traditional or 'non-scientific' medical systems and their potential contributions to health are discussed in the concluding sections of Chapter 2.

The final chapter of the book considers the planning of health services. It is not provided merely as an addendum; rather as an all-pervasive chapter which could equally have been placed first. Planning, it should be realized, may be public and explicit or it may be carried out surreptitiously or without overall aims. However, the increasing importance and spiralling costs of health care have resulted in greater levels of national and even international involvement in health care planning, making it imperative that all those concerned with health care, whatever their specific interest, at least appreciate the complexity of these planning processes. The World Health Organization (WHO) is a key body in international co-ordination, and frequent reference is made to its activities and projects. However, there are many problems associated with co-ordinating international and national plans for health care. These are also manifest between the national and local levels, the latter being the level at which the realities of health care (or the lack of care) become more apparent. Scale of analysis and questions of congruence amongst international, national and local plans therefore form important components of Chapter 8, and the last-named topic, in particular, is receiving increasing attention in planning today (Eyles et al. 1982; Grime and Whitelegg 1982; Phillips and Court 1982).

Medical geography

The subject matter of this book is intrinsically concerned with a long-established area of spatial study, medical geography. This has been variously defined, earlier definitions in particular emphasizing its interest in the incidence of diseases, the distribution of physiological traits in different communities, and the correlation of such data with features of the natural environment (Howe 1972). This focus in medical geography on the correlation of diseases and disease distributions with possible or actual environmental causative factors is still strong and worth while. This type of geographical study has been advanced by a number of authors in the English-speaking world (Howe 1972; McGlashan 1972; Learmonth 1978), although particular reference should be made to the work of Jacques May, a French surgeon, who developed the concept of disease ecology, seeking to explain the prevalence of disease by reference to geographical-environmental factors called 'geogens' (May 1950, 1958).

'Disease geography' (nosogeography) has also been extensively reported in the non-English academic literature. Indeed, in countries such as the USSR, it seems that it is only this type of academic medical geography that is permitted and that health care planning, elsewhere in part a geographical interest, is left in the hands of the State (Learmonth 1978). In France, Belgium and Germany, this traditional medical geography has proved particularly strong. It has received general attention for many years (Sorre 1933, 1943, 1947, 1966; George 1959, 1978; Picheral 1976, 1982), in demographic terms (Beaujeu-Garnier 1966), in studies of specific diseases (Verhasselt 1975), and in analyses of patterns of disease, mortality and health care (Picheral 1976; Diesfeld and Hecklau 1978). The Spanish- and Portuguese-speaking countries have also been active, mainly in disease ecology but with some medical-social studies developing (Arroz 1979; Kroeger 1982; Villar 1982). It certainly repays the student of medical geography to obtain some insight into the vast amount of published work which is available in languages other than English although, because of the amount of this alone, for ease of reference, the examples in this book have been drawn mainly from publications in English.

Learmonth (1972, 1978) deals in detail with the concepts of disease ecology, disease agent, host and environment relationships, and what has been generally termed 'ecological medical geography'. This may be regarded as a traditional and well-established focus of medical geography (Phillips 1981a; Howe and Phillips 1983). Within the geography of disease there are a number of important concerns, including the ecology and environmental associations of specific conditions (see, for instance, entries in Howe 1977, and McGlashan and Blunden 1983), their statistical and cartographic analysis, perhaps modelling spatial diffusion or temporal variations (Murray and Cliff 1977; Beaumont and Pike 1983; Haynes 1983), and the identification and description of environments which are in some ways unhealthy (Howe 1982). This last-named concern is a direct development from medical geography in the days of exploration when areas dangerous to health were identified. With the aid of WHO data and modern cartographic techniques, much more sophisticated and informative description can be carried out, helpful to preventive medicine in particular.

Ecological medical geography has been supplemented in recent years by a very rapidly growing school-within-a-school. Giggs (1979) has usefully distinguished a threefold categorization of spatial research into human health problems: the spatial patterning of ill-health and mortality; the spatial patterning of the physical and human environmental characteristics which adversely affect man's state of health; and the spatial patterning and use of the main elements of health care delivery systems developed to combat diseases and the environmental hazards which adversely affect man's health. The first two research interests are clearly related to the disease ecology or traditional thrust in medical geography. The last-named comprises more of what may be termed a 'contemporary approach' in medical geography. Phillips (1981a) suggests that this contemporary, medical-social geography involves research into the location, planning, and utilization of health care facilities, together with the identification of those features of health care delivery systems that influence their efficiency and effectiveness.

It is with this side of medical geography that this book is mainly concerned: the spatial characteristics of health care systems, the nature of populations served, and the eventual use (or non-use) of facilities. In particular, the influence of decision-makers such as politicians or health services planners and managers on the use of services is considered important, and forms a key area of managerialist research to date (Pahl 1970; Herbert 1979; Phillips 1981a). The activities of managers and professionals who organize and arrange systems can essentially amount to a rationing of health care in any country. This is because demand for health care will almost always exceed the supply capacity which can be provided, necessitating rationing either by waiting lists, queues, pricing policies, underprovision, regulative restrictions, or by some combination of these (Phillips 1981a). As Knox (1982a) suggests, bureaucrats often rule the systems!

There is an element of overoptimism in the third of Giggs' themes mentioned above. The twin poles of medical geography have rarely been effectively co-ordinated or merged (Howe and Phillips 1983). Indeed, it is still possible to agree at least partly with Learmonth (1978) when he suggests that there may be two medical geographies rather than one.

In recent years, most spatial research in contemporary medical geography has concentrated either on describing and optimizing distributions of health care facilities or on describing and modelling their use. Only infrequently have the distributions of disease-causing agents or of disease been taken into account. One of the earliest studies to do this is still, perhaps, the most sophisticated attempt to model changing disease distribution patterns and associated changes in needs for caring facilities (Pyle 1971). However, the complexity of variables underlying the changing nature of disease patterns and the utilization of facilities is such that medical geography is still at a relatively infant stage in conjoining the two branches of the subject. As Shannon (1980), Phillips (1981a), Mayer (1982), and others suggest, there is much room and great need to combine effectively the traditional disease ecology approach with that of the health care geographers.

'Contemporary' medical geography has grown in association with recent changes in human geography. The 'quantitative revolution' and the development of a welfare orientation have influenced its characteristics, as have recent research foci such as managerialism and political economy within welfare geography (Norman 1975; Harvey 1983). Phillips (1981a) has highlighted 'contemporary issues' in medical geography, discussing in particular the changing nature of health care and associated organizational and accessibility changes. Other writers follow the nature of medical geography as it relates to health and to social geography as a whole (Eyles and Woods 1983), whilst Whitelegg (1982) has focused on inequalities in health care and on consequent problems of access and provision, which perhaps accords more with the approach adopted here.

In recent years a number of academics have been at pains to point out that medical geography has become too broad and, perhaps in itself, of less relevance than would be a 'geography of health' (Garcia 1975; Picheral 1982). A geography of health would avoid perpetuating an unrealistic and artificial dichotomy between the geography of disease and the geography of health care. It would, ideally, combine the spatial analysis of health phenomena and health care facilities and would have relevance as much in developing as in developed countries. However, this state of the art has as yet hardly been reached and the causes of this lie not only in the internal divisions within medical geography but also in serious practical impediments to conjunction. A lack of data on morbidity at suitable spatial scales is a frequent hindrance to the development of, for example, intraurban health care locational strategies which can take account of disease patterns. In addition, there are very real problems of forecasting spatial demand for medical facilities in isolation from other societal and economic trends. Finally, a major stumbling block to the practical integration of disease and health care geography remains one of poor understanding of the aetiology (causation) of many diseases, particularly chronic ailments and mental illness (Giggs 1983a, 1983b; Giggs and Mather 1983).

The possible conjunction of these two branches of medical geography is discussed further in Chapter 8, but it is as yet still far from being generally achieved. Indeed, we are conscious in this volume that we might be accused of maintaining an artificial dichotomy by focusing on health care accessibility and utilization (for example, in Chapters 2 and 6) without clear reference to levels of health and patterns of disease in the examples introduced. It is certainly not our intention to foster this division, and we would hope that this volume, in advocating an interdisciplinary approach and recognizing spatial contributions and limitations, may highlight the fact that internal divisions within medical geography itself should become of little relevance in the wider sphere of research. In future, perhaps, they will fade away. In this case, a geography of health might indeed emerge, although no apology is offered for currently focusing on only part of it.

Disciplinary perspectives in health care

It has been suggested that a major problem in health services (and health) research may well reside in the 'disciplinary' research approach. 'Health and health services research is, in fact, not a scientific discipline; rather it is a pragmatically oriented, problem solving activity' (Shannon 1980, p.1). This is probably a justifiable assessment but, for the intending student of medical geography, it is essential to achieve some understanding of the disciplines which contribute to health and health services research or otherwise they will tend to remain isolated and their contribution will be restricted. Therefore a synopsis is now provided of the major disciplines, other than geography, engaged in cognate health research although, at this stage, all possible links and interactions are not explored.

A useful starting point is the implicit classification contained in the journal *Social Science and Medicine*. Until 1981, it was published in six major parts (which had grown over the years): medical sociology; medical anthropology; medical economics; medical geography; medical psychology; and medical and social ethics. Subsequently, the journal has been published as one volume whilst maintaining the same internal divisions, except for medical economics being retitled health economics, and the addition of a section on health policy. The major social science disciplines and topics are identified in these broad headings.

In practical terms, an important divide appears to be between the work of medical and biomedical sciences and that of social science. A few geographers and some social scientists have linked and worked effectively with medical researchers but, generally, this is not common and the contribution by doctors to geography (and also to social science) has on the whole been greater than the reverse (Learmonth 1972). A number of medical specialisms do, indeed, have some explicitly social and spatial orientations. These specialisms overlap and are frequently known by different names internationally but 'community medicine' is perhaps the most general. It is increasingly important in Western medical education and involves aspects of medical practice formerly included under the titles of epidemiology, preventive medicine, and public   health and organization of medical care (Farmer and Miller 1977). 'Medical statistics' and 'operational research' often provide a link between this general discipline and the social sciences, as illustrated elsewhere (Phillips 1981a; Howe and Phillips 1983).

Epidemiology is one of the better-known socially oriented medical services. It has its roots in the nineteenth century, at which time it was primarily concerned with the spread and control of infectious diseases. Today it is a specialty within medicine itself, and, as such, is concerned with the diagnosis and treatment of disease and with health maintenance. Epidemiologists analyse the causes of disease and define methods of disease prevention and in particular they study the frequency of disease, disability, and death in groups of people. The subject is the science of the mass phenomena of diseases, infectious or otherwise and although epidemiologists are increasingly becoming involved with the location and use of the facilities, these activities are less well defined (E.G. Knox 1979; Parry 1979). In contemporary epidemiology there are at least three principal types of investigation, usually known as *descriptive* studies, *analytic* studies, and *intervention* studies. Descriptive studies measure and outline different health problems in communities and attempt to draw inferences about prime causes and determinant factors which correlate with the frequency of disease. Analytical studies usually begin with a specific hypothesis about possible reasons for observed distributions. They aim to explain why some people contract the disease in question and to suggest preventive measures. Intervention studies often introduce preventive or therapeutic measures to alter the distribution or progression of disease. These could be medical intervention, health education, or environmental improvements, and commonly involve the epidemiologist in clinical trials or detailed cross-sectional, cohort, or case studies. These three basic approaches to epidemiology can be further refined by temporal and spatial analysis, which provide a very definite link with medical geography. The epidemiologist, particularly when working as a community physician, often has very broad interests and responsibilities relating to control of infectious disease, public health, and health services, providing yet further avenues for interdisciplinary work. Epidemiologists have close links with paramedical workers and disciplines such as biometry and demography (the study of population movement and change). Today, epidemiology is recognized as having a very important contribution to make in health services planning (E.G. Knox 1979).

Within the social sciences only brief mention need be made of the disciplines identified. *Medical sociology* has, perhaps, one of the broadest orientations although there are specific foci within it. Sociologists were prompted to organize medical sociology as a field of sociological enquiry by the recognition that medical practice represents a distinct segment of society, with its own social institutions, processes, occupations, problems, and behaviour (Cockerham 1978). It has developed strongly as a subdiscipline in North America and Britain, as well as in some European countries, and to some authors, such as McKinlay (1971, 1972), medical sociology can stand in the same relation to medicine as some of the biomedical sciences, physiology, and psychology.

There are, within medical sociology, many well-known writers (Cartwright 1967; Freidson 1970; Freeman et al. 1972; Stacey 1976) and distinct fields of interest. The relationship between social factors, illness, and the behaviour of persons who feel that they are in need of medical attention (a type of 'social epidemiology') is one important focus. Similarly, age and sex differences in morbidity and utilization patterns and the influence of the family and social networks on attitudes to, and use of, health services have been important topics of study. Professional–client relationships and the attitudes, roles, and behaviour of health care professions, insofar as they influence and affect lay persons and the quality or equality of care, have also been extensively researched. Health institutions, hospitals, clinics and health centres are also important foci of research – both their formal institutional characteristics and the effects of their organization on patients, potential patients, and staff. Finally, medical sociologists have explored more general matters of health and society, particularly as these pertain to patients' perceptions of care, ill-health, and death, and the wider questions of the role of health and health care in society. For convenience, although in reality a professional rather than an academic discipline, medical social work should be mentioned here. Although it does not exist in all countries, medical social work tends the social needs of the sick and their families. Because it derives in part from the distribution of charitable money by hospital almoners, it occurs mainly in an institutional context but, increasingly, medical social workers engage in the community, sometimes linking with doctors and nurses in primary care teams (see Chapter 4).

Medical anthropology researches a similar range of topics to medical sociology and it would be surprising if this were not so since many university departments of anthropology and sociology are joint. The focus is often on cross-cultural studies of health, health behaviour, and disease, and these are frequently set in developing countries (Kroeger 1983). Some medical anthropological studies have an historical context: for example, studies of the effects of 'imported' infectious diseases on indigenous people in pre-Spanish conquest South America or the impact of external medical systems on small-scale societies.

Health economics, on the other hand, is a more clearly defined field although also with diverse foci. It is related to the general subject of welfare economics, which deals with the nature of supply and demand, costings, payment, and market forces in welfare. Smith (1977) provides a useful introduction to this theme for geographers, whilst Mishan (1964, 1969) and Nath (1973) provide a welfare economic perspective, and Mandel (1973) and Desai (1974) set out the basics of Marxian economics. Health economists focus the welfare debate on health services, identifying as targets for analysis topics such as consumer choice, elasticity of demand, collective choice, and resource allocation in health services. In particular, differences and potential conflicts between collective choice or good, and individual choice or good are highlighted by what is essentially macro-scale economic analysis. Concepts such as optimality and methods such as cost–benefit analysis are of central concern in health economics. However, the subject also focuses at a more micro-level on, for example, the economic viability or efficiency of specific plans and individual institutions. As Smith (1977) details, there is considerable utility in adopting some of the concepts of welfare economics to examine the distribution of welfare in geographical space. In Chapter 5, it will be evident that this is possible using spatial adaptations of some welfare measures such as the gini coefficient.

Medical psychology deals with attitudinal and perceptual matters relating to health and health care. These include attitudes to facilities and providers as well as matters such as the impact of mass-communication health campaigns and levels of knowledge of specific medical procedures. In particular, there is developing a field of bio-psychology, which has links with biology and medicine, researching, for example, the acceptability of new methods of birth control, attitudes to mental ill-health and its treatment, reasons for smoking or obesity amongst subgroups, and biological bases of intelligence. *Medical and social ethics* is a more diffuse field, researching topics such as the ethical acceptability of new medical treatments and procedures and ethical questions relating to medical research. It also considers the medico-legal implications of developments in health care and of matters such as diagnosis for administrative purposes (Lomas and Berman 1983).

The final subject identified in *Social Science and Medicine – health policy* – is of considerable cross-disciplinary interest. Not only does it involve political decision-makers and health care planners engaged in policy formulation but it also involves economists, sociologists, psychologists, and geographers interested in the implications for their own subject of developments in health policy both nationally and internationally. In Chapter 8 the case is argued for the involvement of medical geographers in interdisciplinary planning; and their involvement in the social and spatial aspects of health policy formulation is equally important, as this is the precursor of planning. Breadth of disciplinary involvement seems to be crucial at the research stage, Kinston (1983), for example, calling for pluralism in the organization of health services research and flexibility with regard to the commissioning and evaluation of research. This is essential if the shallowness and limitations of a single-discipline approach to health services research and planning are to be avoided. However, as Shannon (1980) notes, truly interdisciplinary research efforts are still all too rare.

Geographical perspectives on health care provision

Two important geographical perspectives on health care provision have already been discussed, namely, access (and potential access) and utilization. These, we feel, are two sides of the same coin, as utilization is a manifestation of 'revealed accessibility'. As discussed in Chapter 3, health care utilization and health care need are not, however, identical. Health care is the most difficult service to 'shop' for intelligently (De Vise 1973). Patients or potential patients usually do not know the specific services needed and have little or no knowledge or competence for assessing quality of physician or treatment – indeed, opinions on these things may differ considerably even amongst health care professionals (McKinlay 1972; Ben-Sira 1976). The social position of physicians and the use of medical jargon further widen the gap between professionals and clients and 'need' is very differently perceived and defined by each group. Therefore, it is probably fallacious to argue that where rates of utilization exist, need is greatest. Other measures of health care needs, such as mortality rates, disability days, and life expectancy, may be much better indicators. De Vise (1973) explains that, although poor and minority people in the USA may experience more illness than non-poor and white groups, their utilization of health care services is not correspondingly higher. In fact, their utilization of medical and dental services may be considerably lower than that of better-off citizens. In developing countries, of course, such differences between privileged and underprivileged can become even more exaggerated.

Access and utilization apart, there is a third facet of health care which, in fact, received considerable early attention from geographers. Systems design and optimization have been of interest since Central Place Theory ideas suggested hierarchical arrangements of service centres. Subsequently, public services provision has been subjected to sophisticated modelling and analysis such as spatial interaction modelling or linear programming. Accessibility to services is in part a function of the distribution and location of service facilities, although travel distance per se may be less important than travel time, costs, or convenience (Ingram 1971; Massam 1975; Smith 1977; Moseley 1979). Geographers have, nevertheless, concentrated on assessing the extent to which distribution patterns deviate from the spatially optimal (however defined). This has been done for general public services by Massam (1975, 1980) and Askew (1983); for hospital systems by Schneider (1967), Gould and Leinbach (1966), Morrill and Earickson (1968a), and Morrill and Kelley (1970); and for primary physicians by Curtis (1982) amongst others. All of these optimization applications touch on utilization but usually indirectly. Rather, they are concerned primarily with the mechanics of the supply system. Only a few locational modelling studies, such as that of community hospitals in East Anglia, England, have included behavioural research to assess the potential effects on consumers and suppliers of possible changes in the distribution of facilities (Haynes and Bentham 1979).

Nevertheless, systems optimization is a fair objective, for, as Massam (1980, p.2) asks, 'if you had to find a site for a new school, a hospital, a park or an airport... how would you go about the business of evaluating alternative locations? Is it possible, theoretically and practically, to define the attributes of a best location and is it possible to find the best location?' This is essential information for planners, politicians, and the public at large. However, it is only rarely that a totally new system will be designed, since usually the effects of adding or subtracting relatively minor components are being measured (illustrated in planning terms in Chapter 8). Therefore, the questions of access and utilization may be regarded as being more general and pressing than those related to system optimality, although complementary.

For planning purposes, a great deal more basic information and theory is needed on the variables which influence access and utilization in established systems and about their interlinkages. *This is the emphasis of this book.* What is the point of designing a *spatially* efficient system if the implications for its use are not known? There are numerous personal and institutional variables to distinguish and investigate. For example, the size of a hospital is crucial to its efficient functioning. Technologically, a single giant hospital may best serve a region, but diseconomies of scale in staffing, and inaccessibility for most patients, may well result. However, it may be medically and technologically impossible or undesirable to subdivide the unit into numerous smaller facilities, which would give better spatial evenness of distribution. Therefore, utilization, accessibility, and location may all have to be compromised beyond that which is optimal. The process of spatial search, as Massam (1980) calls it, implies a selection and balance amongst a number of possible strategies. Intelligent spatial search demands an adequate grasp of the relationships between system design and access and utilization.

Overall, therefore, there seems to be a clear and important geographical perspective on health care. The focus here on accessibility and utilization will, we hope, redress to some extent the earlier preponderance of studies attempting to optimize systems with little detailed consideration of how proposed alterations would be implemented or of their implications. As a background to the remainder of the book, a discussion now follows of the nature and implications for accessibility and utilization of various types of national health care systems.

2 Health Care Delivery Systems: some International Comparisons

Factors creating diversity amongst health care systems

As suggested in Chapter 1, there is a range of factors which create diversity amongst health care systems. These can, in very basic terms, be summarized as follows:

1 *The nation's health policy*, which usually reflects its economic and political ideology.
2 *The degree of centralized control* over health services.
3 *Methods of payment* for medical services.
4 *The relationship between physicians and patients*, including the degree of choice and continuity of contact.

Even a cursory glance at these factors will alert the reader to the fact that all are very closely interrelated. For example, the freedom a person has to select and pay for a physician or a certain type of medical service depends very much on the nation's health policy. This, in turn, will be directly influenced by the nation's economic and political orientation and, therefore, it is impossible to escape from the conclusion that the political economy of a nation is fundamental in determining the nature of almost every type of service within it, regardless of whether or not it is provided directly by the State.

A number of authors have focused on the political economy of the welfare state in general (Gough 1979; Wilson and Wilson 1982) and of health in particular (Doyal and Pennell 1979). The very notion 'welfare state' implies that the State has a direct influence in providing welfare services for a nation. It is recognized, however, that the political economy of certain nations, such as the United States, tends to hinder the development of a welfare state. Therefore, government involvement in provision of health and welfare services should not be assumed as the universal norm. The term 'welfare state' itself is not tightly defined but its central concerns are generally accepted to be fourfold. First, there are various monetary social security transfer payments for many types of pensions and sickness and child benefits. Secondly, there are benefits paid in kind in the form of health and personal social services. Thirdly, state-financed education can be provided and, fourthly, there is a variety of miscellaneous subsidies such as of school meals or of housing. There are also less well-defined categories of subsidy such as tax allowances and rebates or grants to agriculture and industry, which make the boundaries of the welfare state hard to delimit with any consistency from one nation to another (Wilson and Wilson 1982).

Comparative health policies

Social scientists have traditionally been concerned with the manner in which wants and needs have been satisfied when the market is the social mechanism in operation. However, when 'wants' and 'needs' are determined and expressed by the political authorities in any given country, researchers are drawn into a study of the political machine, and this is increasingly true in the study of the various components of the welfare state. Foremost amongst these are health services, the levels of capital and current expenditure on which are such that few, if any, national governments have been able to remain totally aloof from intervention and, perhaps, active involvement in their planning and provision.

Even if the State does not directly finance health care, its attitudes will colour the nature of the service available. Licensing of health professionals and of institutions, minimum standards of safety, and attitudes to professional autonomy vary from nation to nation and affect the nature of the service provided. A number of authors have constructed typologies of international health care systems (Babson 1972; Kohn and White 1976; Roemer 1977; Maxwell 1980; Roemer and Roemer 1981). These have generally relied upon existing political and developmental classifications distinguishing, for example, between socialist and capitalist economies and between developed and developing nations. Roemer (1977) has produced a useful general statement that three sets of factors, economic, socio-political and cultural-historical, have influenced the development of international health care systems. He classifies political systems as centralized, moderate or localized structures, and economic systems as affluent, marginal or very poor. As Pyle (1979) points out, within this classification, five 'types' of health care systems can be identified: (1) Free enterprise; (2) Welfare state; (3) Transitional state; (4) Underdeveloped; (5) Socialist.

This typology provides a useful basis for later sections although a note of caution should be sounded in assigning any country to a specific slot. For example, an oversimple distinction is often drawn between the public and private health care models as supposedly exemplified by the health care systems of Britain and the USA. In reality, neither country conforms entirely or strictly to either categorization: there is a small but growing private health service sector in Britain and, in the USA, the Federal and state governments have become increasingly involved in health services and are responsible for over two-fifths of expenditure on health care. Elsewhere, in Europe and in the developing world, there are systems which combine in varying patterns compulsory health insurance, free provision of some services, charges for others, and public and private supply of health facilities (Wilson and Wilson 1982). Therefore, if it is difficult to be precise about the nature of the extremes in the classification, considerable diversity and overlap are to be expected at intermediate positions!

Virtually all nations have adopted some responsibility for health care but there are enormous variations in the extent of financial support and in the ways in which control is implemented (Figure 2.1). Babson (1972) and Maxwell (1980) suggest that there is almost invariably a cross-cultural evolutionary process at work which is tending towards total government control of health care delivery. However, nations are at very different stages along the path, some even witnessing changes in direction; in Britain, for example, increasing levels of privatization have been encouraged by Conservative governments, whilst Labour governments have generally opposed this trend. The diversity in control can be exemplified by hospital organization. Within Europe, Babson (1972) suggested a fourfold categorization of organization types. The Scandinavian systems (Denmark, Finland, Norway and Sweden) are typified by considerable local autonomy; the Eastern European systems found in the USSR, Poland, Hungary, and East Germany, have more centralized ownership and control; the central European types found in Germany, Austria and Switzerland have provincial governments largely responsible for hospital services; and, finally, diversity of ownership and complex controls are to be found in the Latin-Benelux countries such as Italy, Belgium, France and the Netherlands. When Spain, Portugal, Greece and

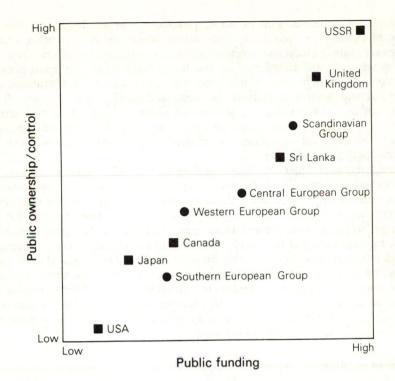

Figure 2.1 The extent of public ownership and control of health services in various countries

Source: after Maxwell (1980)

Turkey are considered, it becomes increasingly difficult to provide any sustainable categorization of hospital systems even for Europe. Therefore, to attempt this on a world scale in anything more than a highly generalized manner would be patently unwise.

It is perhaps best to agree with Eyles and Woods (1983) who argue that the emergence of a 'welfare state' is not an inevitable outcome of modernization or industrialization. Indeed, the specific tensions between state, capital and social groups in different countries make this unlikely and a considerable range of health care systems has emerged in industrialized countries. There may be a distinct, if not universal, trend towards greater government involvement but, perhaps, 'societal constraints' are more important in influencing the nature of health care and the development of health care policy. These include the nature of the containing society, conceptions of illness prevailing and responses to health and illness (Eyles and Woods 1983).

The degree of centralized control

The extent to which control over health services is centralized will tend to influence the sensitivity of a system to local needs. At one extreme, central authorities determine all financial disbursements, work norms for physicians, size and design of institutions, number of patients per physician, and patient allocation. At the other extreme, decisions are made at a local level to meet local needs and there is no overall framework for the system or formal direction of its constituent parts, which can result in overlap of functions, diversity in types of provision, and deficiencies in disadvantaged areas. Professional and institutional standards, costs and quality will all tend to vary considerably and both patients and service can suffer. Between these two extremes must lie compromise positions in which central direction of policy and provision is tempered with sufficient local autonomy in day-to-day administration to enable local needs and conditions to be met.

It is sensible to consider why health services are organized in particular ways. The answer should, perhaps, be: *so that the supply system can meet consumer demand.* In addition, there is the need to make the most effective use of available resources by organizing them and regulating their quality and efficiency. It will be seen that some systems perform better than others on these goals. In some countries, most attention is given to the engineering of the supply system, with norms judged from a managerial and predetermined view. This is a criticism often levelled at socialist and some welfare state systems. In others, notably free-enterprise systems, only those who can afford to pay will generally have access to health care. In effect, this will tend to remove from the most needy any care at all. So, in both extremes, the position of the consumer may be prejudiced by the nature of the delivery system which, in turn, is dictated by national health policy and political economy.

Methods of payment

Methods of payment for medical care vary according to the timing of payment and sources of funds. The timing of payment concerns whether monies are paid when treatment is received, or whether payment is made to a common fund (either to a source of care or to an insurance company) prior to treatment. The objective of most collectivized health services, including the British National Health Service (NHS), is to be 'free at the time of use'. Many countries now provide free, or virtually free, hospital treatment but there may be user charges for other physician and dental services, for medicines and for prosthetic appliances. Economic access barriers have, in many developed countries, been reducing, although, for reasons discussed in Chapter 6, equal economic access to health care still does not mean that rates of receipt are precisely proportional to the needs of the population (Roemer and Roemer 1981).

Funding for medical care may be from a variety of sources. There may be a scheme of national health insurance under which employers and employees provide funds to pay insurance premiums to schemes run by, or on behalf of, the State, which subsequently pays for care of the insured population. Sweden pre-1974, Norway since 1956, the Federal Republic of Germany and Japan are examples of this type of care (Deppe 1977; Roemer and Roemer 1981; Chester and Ichien 1983). National health services, on the other hand, are supported out of collective taxation by central treasuries (which, of course, derive income from employers and employees). Every citizen then has a right of access with or without fees for specific items. This form of central financial provision for health care is still comparatively rare, but it is exemplified by the British NHS, and by the systems in the USSR and China. Canada represents a hybrid of these two forms, with funding coming from provincial insurance schemes (with premiums) *and* from the provincial and Federal treasuries. Finally, individuals may pay privately for care according to their means. If eligible or able to afford the premiums, they may insure against illness and thus have all or part of their costs covered. The USA, for example, has one of the smallest direct government involvements in health service financing and, in spite of the gradual encroachment of public financing into the health care sector, over one-half of the total bill is still paid by various non-government parties (Tripp 1981).

Whatever the source of finance, a major and striking factor is that the costs of health care are escalating almost everywhere, usually ahead of inflation and often ahead of increases in Gross National Product (GNP) – see Chapter 8. Indeed, it is claimed that every country is experiencing an increase in costs as well as benefits, regardless of the particular system of financing (Maxwell 1980; Roemer and Roemer 1981). Most developed countries now spend upwards of 6 percent of GNP on health services; and a major question is for how long this increase can be maintained (Figure 2.2; Table 2.1).

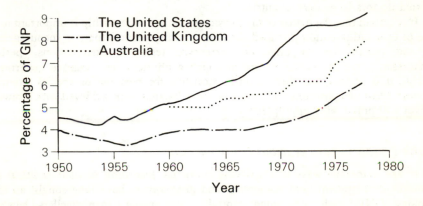

Figure 2.2 Percentage of GNP devoted to health care

Table 2.1 *Health care expenditure as a percentage of GNP in selected countries*

	West Germany	United States of America	Sweden	Nether- lands	France	Canada	Australia	United Kingdom
1950		4.5	3.4		3.4	4.0		3.9
1960		5.3	4.7	4.5	4.7	5.6	5.0	3.8
1965		6.2	5.6	5.3	5.8	6.1	5.2	3.9
1970	6.4	7.6	7.4	6.3	6.4	7.1	5.5	4.3
1975	9.4	8.6	8.5	8.1	7.9	7.1	7.0	5.5

Source: Maxwell (1980)

Physician–patient relationships

A final feature of health care systems to be noted also varies according to the political economy in a given country, namely, the nature of the relationship between the patient and the professional. At least three types of relationship are possible (Shannon and Dever 1974). These have different financial, professional and medical implications. The physician may be a free practitioner, effectively entering into a private contract with each patient for an item or course of care. This occurs most frequently in the USA but numerous other systems have private practitioners who can work in this way. Secondly, contract physicians undertake to treat an association of consumers who prepay costs by subscriptions. In pre-NHS Britain, the so-called 'board' doctors attended a 'panel' of patients who paid a certain sum regularly to ensure treatment. Again, such systems of prepayment are fairly common in the USA. Finally, doctors may be officers of a governing authority, national or local, retained to treat either patients in a specific institution or those designated to be tended by the employing authority. This can include doctors retained by companies to attend employees, and salaried doctors in socialist countries.

It is certain that the nature of the physician–patient relationship is important to the type of care and attention provided. In some private systems, there is encouragement for the practitioner to undertake extensive tests or examinations, perhaps unnecessarily, whilst, in the system in which a physician is a salaried government official, it is argued that there is little incentive for personal or special attention. Between these two positions there must be some balance to be achieved which rewards patient and practitioner appropriately.

Health systems in various countries – some examples

There are numerous ways of classifying countries according to their political and economic development, and economists and geographers have long sought an ideal typology. GNP, birth rates, infant mortality, percentage urban dwellers, levels of literacy and industrial mix can all be used to provide a variety of rankings (Todaro 1977; Mountjoy 1978; Dadzie 1980). In addition, many terms can be used to distinguish countries at different levels of development, for example, 'industrial and developed economies', 'postindustrial economies', or 'developing nations' or 'the Third World' (this last term being preferred by some of the nations concerned). The World Bank (1981) classifies countries as follows (the number of countries in each class is given in parentheses):

Low-income countries	(36)
Middle-income countries: oil importers and oil exporters	(50)
Industrial market economies	(18)
Capital-surplus oil exporters	(4)
Non-market industrial economies	(6)

This classification is useful and, in conjunction with Roemer's classification of health care systems discussed earlier, provides a basis for discussing examples of the spatial and administrative structure of selected countries in each group. Owing to space limitations, these will be rather brief synopses and, in cases, oversimplifications. However, the literature cited will provide further reading on specific systems.

Health care organization in developed countries

The United Kingdom

One of the most widely researched and best documented health care systems in the world is the National Health Service (NHS) which serves the United Kingdom's 56 million people. It grew from a recognition (which matured in the 1930s) that access to health care was not equal for all, and neither were resources evenly or equitably distributed. The NHS Bill became law in 1946 and the NHS began existence in 1948, the dates being of significance in that the NHS was seen as part of the new social order in postwar Europe.

The Act was built on four important principles, which remain substantially unchanged today. The first was *comprehensiveness,* in that the NHS aimed to meet all recognized medical needs and conditions, acute and chronic – an open-ended commitment which has increased as medical science has developed. The second principle was *universality:* that the service was to be available free to all residents and bona fide visitors without further qualification. Thirdly, there was to be *collective financing* from general taxation rather than charges to users and, finally, there was to be *professional independence* for the medical groups in the service.

The NHS did not, at first, overcome the hospital orientation of British medical care that had been growing steadily throughout the twentieth century. Indeed, general practitioners (GPs) often found little professional or financial reward in the NHS. However, from the late 1950s the 'renaissance of general practice' occurred (Hunt 1957). This was partly the result of changing attitudes, which allowed that GPs could be good doctors! It also stemmed in part from changing patterns of disease away from acute, infectious conditions as the main causes of morbidity and mortality towards chronic conditions such as heart disease, cancer and stroke. These conditions may require periodic hospitalization but they can also be dealt with adequately in the community, as discussed further in Chapters 3 and 8.

The NHS as established in 1948 represented very much of a compromise between what was desired and what was achievable given financial and professional constraints (Office of Health Economics 1974). The administrative structure which existed for the first 25 years of the NHS until 1974 very much reflected this and, from the outset, it was regarded as rather unsatisfactory. A 'tripartite system' under the Ministry of Health (which was later to join with another department to become the Department of Health and Social Security) divided responsibility for health care. Hospital authorities (15 regional and 36 teaching hospital boards) had responsibility for specialist and hospital services. Public and community health, preventive and ambulance services, health centres and various community nursing services were the responsibility of local government, with 174 local health authorities existing by 1974. General medical services – doctors, dentists, chemists (pharmacists) and opticians – were separately administered by 134 executive councils in England and Wales. This threefold division resulted in poor liaison, replication of services, and inefficiency; and usually the patient was the loser in quality and continuity of care.

As a *spatial* planning system, the tripartite structure had also been unsatisfactory. Hospitals, for example, were planned as institutions on a regional scale and there were sometimes rather poor underlying reasons for the choice of regional boundaries. Local authorities providing health services were very mixed in their sizes and abilities to perform functions. The 174 local health authorities ranged in population from about

60,000 to over 2 million and hence were an ill-assorted group. Likewise, the executive councils covered a comparable range of populations although many dealt with more than one local authority unit, which hampered liaison and planning.

Pressure for some form of structural reorganization and integration of these three branches of the NHS began to mount in the early 1960s. The Porritt Report in 1962 and the paper of intent by the Labour Party in 1968 indicated the desire for change, supported by a subsequent Conservative government, so that, with the NHS Reorganization Act of 1973, the scene was set for NHS administrative reform concurrent with the reorganization of Local Government implemented on 1 April 1974. It was hoped that a unified structure would replace much of what had previously been called 'muddling through' in health care planning in England and Wales (Maddox 1971).

The integrated system sought to achieve a sound management structure with clear definition and allocation of responsibilities and with a maximum delegation downwards, matched by accountability upwards (Office of Health Economics 1974, 1977). Therefore, the basic aim of centralized planning and control with decentralized execution was sought. To achieve this, a three-tier system was created under the Department of Health and Social Security (DHSS). A regional tier was retained with 14 Regional Health Authorities (RHAs) in England, having responsibility for strategic planning and resource allocation at a general level. Below these were what were originally viewed as the 'key' administrative units, the Area Health Authorities (AHAs), given statutory responsibility for running the health service at a local level. The 90 AHAs in England, 8 in Wales and 15 Area Boards in Scotland employed most of the staff and carried out most day-to-day work and administration, mainly at the equivalent of a county level. The professional contractors such as doctors and dentists were employed at this level, executive council functions being taken over by Family Practitioner Committees (FPCs). However, although in 1974 this Area tier was regarded as fundamental (perhaps the role of the region was more in question), in the late 1970s there was yet another reorganization, aimed ostensibly at saving administrative costs.

The AHA level was abolished with effect from 1 April 1982, its functions being devolved to the most local level created in 1974, the District Health Authorities (DHAs). In Scotland, the existing boards continued although 3 agreed to be dissolved, the future of others being still in doubt. Wales retained 8 districts and England some 192 (Figure 2.3). The populations were still far from uniform, ranging from fewer than 100,000 to over 520,000 persons, although the majority were between 200,000 and 300,000. The boundaries of most were those of existing (1974) DHAs, but some replaced whole AHAs. London 'grew' from having 16 AHAs as main management units to have some 60 district authorities, and it is possible that this fragmentation of roles may create problems for the future. However, as will be discussed in Chapter 8, this reorganization is very much in keeping with a move towards more local planning and involvement in the NHS in Britain. Whereas the DHAs in 1974 were regarded as the 'natural' units for health care, the smallest units for which the full range of general health and social services could be provided, they are now also the most important day-to-day working level of the NHS.

An important innovation at the district level since 1974 has been the official inclusion into the NHS of consumer representatives (Office of Health Economics 1974, 1977; Hallas 1976). Before 1974, the consumer was assumed to be represented

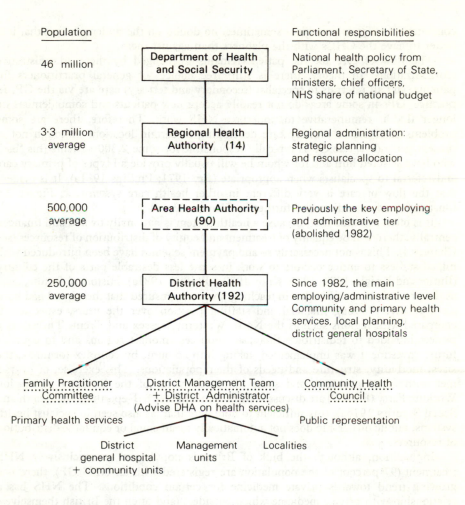

Population

46 million

Department of Health and Social Security

Functional responsibilities

National health policy from Parliament. Secretary of State, ministers, chief officers, NHS share of national budget

3·3 million average

Regional Health Authority (14)

Regional administration: strategic planning and resource allocation

500,000 average

Area Health Authority (90)

Previously the key employing and administrative tier (abolished 1982)

250,000 average

District Health Authority (192)

Since 1982, the main employing/administrative level Community and primary health services, local planning, district general hospitals

Family Practitioner Committee

District Management Team + District Administrator (Advise DHA on health services)

Community Health Council

Primary health services

Public representation

District general hospital + community units

Management units

Localities

( ) no. of units in England

Figure 2.3 The structure of the National Health Service in England since 1974 (as modified in 1982)

in local authorities and by the various lay members of hospital boards. This had resulted in very patchy representation but, today, Community Health Councils (CHCs) are designed to act as public 'watchdogs'. They meet regularly with health service planners and have a duty to visit NHS establishments and to represent the patient and consumer. They can be up to 20 members in size (1982) and, although the levels of co-operation between CHCs and administrators do vary nationally, in many districts, the CHC members bring valuable expertise, gained by visiting institutions, and participate in meetings with an outsider's eye. Many district administrators therefore seek to co-operate closely with their CHCs and to involve them in consultation as far as possible – sometimes, no doubt, on the understanding that it is better to have the CHCs with the planners than against them.

The flow of care for the patient remains unchanged by these administrative reorganizations. In theory, there is still 'free choice' of general practitioners for patients, and all referrals to specialist (secondary and tertiary) care are via the GP. In practice, GPs in some areas do not readily accept new patients and some dentists no longer find it remunerative to undertake NHS work. Therefore, there are some problems of access to primary care developing in certain locations although not at present generally. The GP is typically responsible for some 2,300 patients (his 'list') who have selected him and for whom he will usually provide all types of primary care and referral to specialists when appropriate (Fry 1971; Phillips 1981a). It is evident that the flow of care is very different in other health care systems, as Figure 2.4 illustrates, and as is discussed further in Chapter 3.

It is often assumed that, because a health system is nationally owned and financed centrally, there will be equality of treatment and equity of distribution of resources (see Chapter 4). This is not necessarily so and payment schemes have been introduced with mixed success to entice doctors to work in some less desirable parts of the country (Butler and Knight 1974, 1976; Knox 1979a; Phillips 1981a). Historical factors, such as the pre-eminence of London in teaching hospitals, ensured that the capital had built up a relatively strong financial and staffing position over the years, especially if compared with regions such as the North Western, Wessex and Trent. Therefore, a project intended to redistribute financial resources amongst regions and to equalize future investment was implemented, taking into account, by complex formulae, the sizes, morbidity, structure and needs of their populations. This exercise, of course, met with many problems and some of the activities of the Resource Allocation Working Party (RAWP) are discussed further in Chapter 8 (Department of Health and Social Security 1976a). It is sufficient to note here that, as also seen in socialist health systems, public ownership does not automatically mean equal or equitable distribution of resources.

In addition, although the bulk of Britain's population use exclusively NHS treatment (97 percent of the population are registered with an NHS GP), there is a growing trend towards private medicine for certain conditions. The NHS has a relationship with private medicine which outsiders (and often the British themselves) find hard to understand. Hospital specialists, for example, usually work both for the NHS *and* privately. They may see patients for *paid* consultations and subsequently admit them to *free* NHS wards for diagnosis or treatment (thereby jumping any queue). Indeed, queuing for medical care on long waiting lists is a form of rationing health care in a free system in which demands do not seem to be finite. In the USA, for example, queuing usually only applies to people too poor to attend private practitioners and reliant by default on public hospitals. In the NHS, waiting lists for outpatient appointments are almost a norm of the service.

Therefore, the private sector has grown considerably, especially since 1979, although in the short run its effects should be marginal as its expenditure is only about 2 percent of that of the NHS. By 1980, some 3.7 million persons were covered by private health insurance, an increase of some 27 percent in that year alone, and now include many working-class families. Although still relatively small, the growth of private medicine, some feel, is undermining the foundation of the NHS, the right to health care regardless of ability to pay. It may also restrict the choice available to the

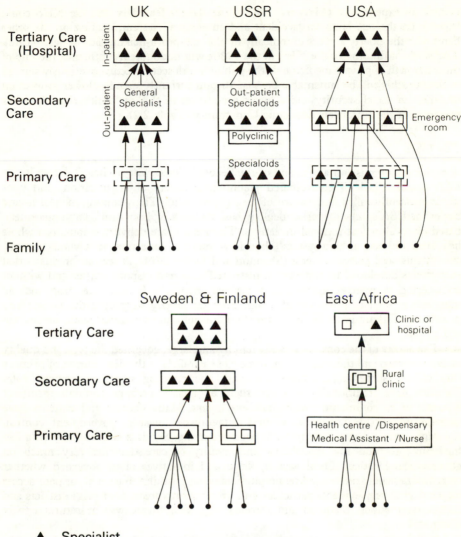

- ▲   Specialist
- ☐   General practitioner/Primary physician
- •   Patient

Figure 2.4   The flow of health care in selected systems

Source: after Vuori (1982a)

less well-off and, perhaps, the removal of the wealthier from reliance on the NHS will reduce it to a second-class service (Politics of Health Group 1982). Private insurance frequently excludes chronic ailments or the problems of the elderly, and the range of activities of the private sector is very narrow: about 30 procedures account for 60 percent of expenditure (Maynard 1983). The danger feared is that the NHS could become on the one hand an 'accident and emergency' system, dealing on the other hand with the elderly and the chronically ill (the non-profit-making side of health care). The levels of funding of the NHS in the 1980s will undoubtedly influence the rate of future growth of private medicine. If the public health sector becomes seriously starved of finance, this will be lamentable because Britain currently spends a lower proportion of GDP on social security and a lower proportion of this on health than do other members of the European Economic Community (Table 2.2).

The USA

A national Commission at the end of the Second World War on hospital care in the USA found a lack of co-ordinated planning, a variety of organization, and poor supervision of quality and nature of care (Pyle 1979). The planning of health care under the USA's free-enterprise ideology has, historically, been limited and piecemeal, based on local or institutional initiatives. The health care delivery system as a whole has been dubbed a 'non-system', but it is more a collection of several systems, subsystems, and partial systems (Shannon and Dever 1974). A reason for this is that subsystems developed in response to restricted aims and responsibilities and without considering themselves as part of a national network. Indeed, the profit-making motivation of many providers of health care has largely supported the laissez-faire principles of minimum official intervention and minimum national planning and co-ordination.

The ability of the consumer to pay has, by and large, governed the type and quality of care received and there have been some close parallels in the distribution of primary care offices (surgeries), in particular, with the retailing system (Thomas 1976). Families have traditionally arranged their own care, perhaps through insurance schemes or in conjunction with their employers. Many doctors and patients have become inherently suspicious of, and antagonistic towards, government control, although some changes have recently been occurring. Services are generally on a fee-for-item basis and the availability and quality of care depends very much on geographical location. Local wealth, drive and initiatives often governed whether communities could attract and retain physicians, and maldistribution of, or poor access to, services is very evident, particularly in urban ghetto areas, poor rural districts and Indian reservations (Shannon and Dever 1974). This theme will be returned to in Chapter 4.

In the past, control of hospitals has rested mainly with government (Federal, state and local), voluntary and religious groups, and proprietary units relying on profits. Regionally, hospitals have varied in quantity and quality and in the types of professionals employed or patients treated. As a result, financial, racial and religious barriers to care exist in some institutions (Morrill and Earickson 1968a; Earickson 1970). Recent research has illuminated the development of the urban medical care delivery system in particular and has identified that the major foundations were laid during the period of urban transformation that took place in the USA during the latter part of the nineteenth century (Knox et al. 1983). The settings of medical care and their relative accessibility to subgroups of the population changed, and the overall dynamics of urbanization, immigration, industrialization and segregation became the key aspects controlling access to services. This may be more important to the structure

Table 2.2 *Comparisons of spending on social security in the European Economic Community*

	Belgium	Denmark	Germany	France	Ireland	Italy	Luxemburg	Netherlands	Britain
A									
Health	22.5	26.8	29.8	26.2	36.3	23.2	23.6	29.3	21.6
Old age	25.8	35.1	25.8	34.9	27.0	34.0	31.1	27.9	40.4
Family	11.6	10.0	8.1	12.5	8.9	7.4	7.9	9.2	11.5
Unemployment	10.4	11.9	3.7	6.4	8.2	1.9	2.1	6.3	8.6
Other	29.7	16.2	32.6	19.9	19.6	33.5	35.3	27.3	17.9
B									
1970	18.5	19.6	21.4	19.2	13.2	18.4	16.4	20.8	15.9
1975	24.5	25.8	27.8	22.9	19.4	22.6	22.4	28.1	19.5
1980	27.7	28.0	28.3	25.8	22.0	22.8	26.5	30.7	21.4

A   Functions of social security spending as percentage of whole, 1980
B   Social security spending as a percentage of GDP
Note: Methods of financing social security are not uniform in the EEC member countries

Source: Commission of the European Communities (1983)

of the present system than the influence of scientific and technological developments or of key professionals and organizations, a theme developed in Chapter 3.

A number of features characterize this mixed system today. Care by specialists extends from primary contact onwards although currently some interest has been shown in introducing a British-style 'family doctor'. Patients can choose a physician according to their means and this influences the flow of care identified in Figure 2.4. Administrative and organizational controls involved a mixture of Federal and state licensing laws, some affecting public health and welfare services, others private organizations. States require hospital licensing but standards have tended to vary considerably. There is no universal coverage under a national health scheme but some categories (noted below) are now eligible for health care from public funds.

The foregoing discussion implies that there has been minimal government involvement in the structure and delivery of health care in the USA. However, since the Depression and World War II, a number of public laws affecting health care have been enacted, the most notable being the Hill-Burton Act of 1946. This Act was designed to supply aid in constructing hospitals in areas of greatest need, a programme which expanded over time and lasted some two decades. This started the process of states gathering systematic data and initiated statewide planning of hospital facilities and the establishing of standards of need. A better distribution of facilities, some higher standards, and co-operation between government and voluntary health care agencies resulted, with improvements coming about in medical care in certain low-income states and rural areas (Pyle 1979). At least the bastion of laissez-faire was breached.

An increasingly complex hospital system was becoming established with a proliferation of health planning agencies, many metropolitan-oriented. By the early 1960s, tentative forms of areawide planning were being sought, some agencies being official, others voluntary, although plans and data collection were not standardized. However, in what Pyle (1979) terms the massive 'barrage' of social legislation of the mid-1960s came two important measures designed to improve health directly. In 1965, Social Security Amendments Title XIX (Medicaid) and Title XVIII (Medicare) provided, respectively, a minimum level of comprehensive medical care to all persons receiving public funds for support and those indigent as a result of catastrophic illness, and partial coverage of medical expenses for the elderly (Kohn and White 1976). These worthy programmes contributed substantially to the need for enhanced overall health care planning.

Other legislation, such as the short-lived Regional Medical Programs aimed to develop programmes to combat heart disease, cancer and stroke, was followed in 1974 by the National Health Planning and Resources Development Act, which was intended to update existing programmes and, particularly, to revise programmes for the construction and modernization of facilities (that is, to replace the outdated Hill-Burton scheme). The programme in effect called for a network of Health Systems Agencies to assist in planning and development nationwide and to gather and analyse health planning data. Therefore, some movement towards co-ordinated planning was evident which, eventually, could have contributed to a national health system had the political will continued.

Government schemes apart, there exist a number of methods of payment and provision of medical care. Prepaid group practice is an important form, in which members of organizations prepay specific fees at regular intervals, regardless of what is used. Kaiser-Permanente is amongst the largest, working with decentralized management nationally. These schemes generally provide for a capitation fee membership and comprehensive coverage (Shannon and Dever 1974). Health Maintenance Organization (HMO) schemes are outgrowths of this trend, again on a prepaid basis for an enrolled membership. Finally, Health Care Corporations (HCC) aim at a centralized control with decentralized execution of medical care and, perhaps, these could bring cost-effective health care to many parts of the country and to underprivileged groups in particular. Shannon and Dever (1974) however, felt that the American Medical Association would not support these on the philosophical grounds that it opposes a rigid and centrally organized health care system and favours pluralistic, private schemes. Nevertheless, there are limits to the extent to which competition is favoured as it could eventually reduce the incomes of physicians (Reinhardt 1982), even if some attributes of competition are very attractive to US attitudes (Enthoven 1980).

In spite of some apparently unifying trends, there is still a great diversity of health care structures in the USA and considerable spatial and social inequalities of access. In particular, many people rely on emergency room treatment and lack a regular source of care. Traditional views and antagonisms to centralized control have been slow to recede but, as in other countries, costs of medical care are spiralling. Indeed, the proportion of GNP spent on health care in the USA has risen from 4.5 percent in 1950 to over 8.5 percent in 1976. The government paid 20 percent of the cost for personal health care in 1950 but by 1975 this proportion had doubled. Over the same time period, direct payments fell from 68 percent to 34 percent. Therefore, whilst government financial commitment doubled, direct (personal) payments halved. Charity payments have also fallen, from 3 percent to 1 percent (Pyle 1979). Personal insurance has increased its importance but the costs of premiums have risen beyond levels which can comfortably be afforded by many.

The reasons for cost increases, a major underlying factor in the crisis in health care provision in the USA (and in other developed countries), are manifold and include technological and administrative complexities as well as inflation (see Chapter 8). In addition, national programmes which could reduce the burdens on individuals are hampered by the ideologically based opposition of certain political groups and by provider groups who perceive threats to their traditional independence. It is also suggested that there are too few doctors to provide the levels of care needed. However, when manpower levels in many countries are compared, the USA does appear to be well provided. Therefore, for many, expectations must have risen beyond a realistic point and whether such health care demands can continue to be met in the future is very much a matter of debate.

Australia

Australia provides an example of a nation which has come to expect a high standard of living for most of its 14.3 million inhabitants, and the decade of the 1970s saw considerable change and experiments in search of a suitable health care system to complement this aspiration. The six states of Australia and the Northern Territory are quite autonomous in their health policies although considerable similarities have developed amongst them. The following remarks are therefore of a general nature but will apply in most states.

L. I. N. E.
THE MARKLAND LIBRARY
STAND PARK RD., LIVERPOOL, L16 9JD

Essentially, primary and specialist care have remained largely private since the turn of the century. Hospital services, on the other hand, have increasingly become state-funded. Today, some 30 percent of the 28,000 doctors in Australia are full-time public employees. Most of the rest are in private practice, charging fees per item of service. As a whole, Australia spends about 9 percent of its GDP on health care.

Up to 1945, the financial core of the Australian health care system involved third-party benefits from various sources of private insurance and government subsidies. In 1945, the Commonwealth (Federal) government offered states a subsidy for public hospital patients and, since 1950, certain medicines have been free. Additionally, pensioners have received free primary care since 1951. A three-tier client hierarchy emerged, with private, fee-paying patients admitted to hospitals by their own doctors, mainly to private rooms. At the bottom of the hierarchy were 'public' patients (means-tested after 1952 except in Queensland). They could not specify which doctors would treat them and were accommodated mainly in large wards. In between the two levels were patients who chose to be treated by their own doctor but usually in two-bed rooms, paying hospital and doctor separately (Goldstein 1982).

Primary care also had a three-tier hierarchy of clients: those receiving free treatment (such as the unemployed or pensioners), those with health insurance, and those without insurance but ineligible for free treatment. The first channel for help was the GP, with subsequent referral to specialists. All those except the 'free treatment' patients paid a fee for service. This system, largely dependent on private expenditure, existed until 1975, although the importance of personal contributions fell from about one-third of medical expenditure in 1970 to one-fifth in 1975.

Goldstein (1982) notes a number of problems in spite of the high standards of health care in Australia. These relate particularly to tax concessions on insurance premiums which did not fall equitably on all concerned. The stigma of a means test for free or subsidized treatment (except in Queensland) was somewhat unpalatable. This, together with the variety of benefit scales offered by health insurance organizations, the three classes of patients and the dual system of private and public health made the Australian health service very unwieldy. Perhaps most worrying was that 10–15 percent of the population, mainly on low incomes, were neither insured nor eligible for free treatment, and medical care costs were rising ahead of inflation. There was a resultant overuse by the insured and underuse by the uninsured, which again caused concern.

As a result, the Federal Labour Government (1972–1975) introduced a publicly financed universal health insurance scheme, Medibank, from 1 July 1975. Under this scheme, public hospital treatment was free and public revenues met 85 percent of medical costs, with a maximum of $A5 payment by the consumer (WHO 1980; Stimson 1981). Pensioners and the indigent continued to receive free treatment. The costs of private health insurance tumbled and subsequent yearly changes to dismantle Medibank by the Liberal-National Country Party Government since 1975 have not stopped the drift from private health insurance. At present, public wards are still free but the patient pays the first $A20 of any medical procedure (Australian Bureau of Statistics 1981). The Australian health system up until early 1983 is therefore only marginally different from that replaced by Medibank, having swung from a system relying mainly on private provision to public care, and now back to private provision (Eyles and Woods 1983). Queensland alone retains free inpatient and outpatient treatment. In the rest of Australia, the poor, the unemployed and pensioners must identify themselves with a card for free treatment – Pensioner Health Benefits (PHB) cards. The remainder are either insured or risk paying the full subsidized cost.

The costs to the government of health care have been reduced by means of these changes and some 59 percent of all possible contributors in 1980 had some type of private health insurance, with 16 percent covered by Commonwealth health benefits. However, 25 percent have neither insurance nor access to free benefits (Australian Bureau of Statistics 1981). Differences have emerged among the states and this in itself may exacerbate inequalities (Goldstein 1982). As in the USA, professional interests are important and the Australian Medical Association acts as a powerful political lobby to protect the free-enterprise system in medical services (Hetzel 1980). Governments institutions, medical professionals and public all seem to have different ideas about what constitutes the right system for Australia and recent changes do not appear on the whole to have been to the benefit of consumers but rather to have produced confusion and generated some resentment towards the political parties' manipulation of health care.

In spatial terms, access to health care has been generally good for Australia's largely urbanized population. There are, however, distinct patterns of 'doctor-rich' and 'doctor-poor' areas in cities such as Adelaide, Melbourne and Sydney (see Chapter 4). In particular, high-status, older-established areas have tended to have the best population:doctor ratios, and the growth of group practices and decline of solo practices noted in Britain is evident (Phillips 1981a; Stimson 1981). Nevertheless, this ratio is generally favourable at around 1 GP per 2,000 patients though there are extreme variations in some city suburbs (Stimson 1981). There are other, better-known problems of access to health care in this vast continent, and people living in isolated areas are readily identifiable as a disadvantaged group. Many of these are served by the Royal Flying Doctor Service, established in 1928, which now has 13 flying doctor bases, and there are also other aero-medical organizations nationwide (McEwin 1981). In addition, the aborigine population suffer disproportionate ill-health and deprivation. Today, there are only 180,000 aborigines and they comprise poor groups in both rural and urban locations although, recently, campaigns have been launched with some success to upgrade their well-being, particularly by using aborigines to manage aboriginal health.

Australia, therefore, does have problems of health delivery in spite of its excellent standards of general care and the vast sums spent on health. To economize on expensive hospital care, some attempts have been made to increase the provision of primary care in the community (Ohlsen et al. 1980). Australians do have good life expectancy but they are not as healthy as their popular image suggests, and neither are their New Zealand neighbours. In New Zealand, there are both public and private hospitals and patients may therefore have a degree of choice within their means (Easton 1980; McEwin 1981). There are, in fact, some notable similarities (but equally, some notable dissimilarities) between the systems in these two countries which repay further study. The Australian system is also in some ways similar to that in the USA, especially in the autonomy afforded to states and in the professional resistance to central control. What direction this will take in the future and the implications for Australian health services are not clear but merit close monitoring.

Some other developed countries

There are, of course, developed countries with health care delivery systems substantially different from those previously discussed and some of these provide very good care. In particular, the Scandinavian countries of Norway, Sweden, Denmark and Finland have distinct similarities in their health service systems. They can be regarded as mainly public, highly structured and socially planned. Physicians are generally salaried and services financed by employee/employer contributions. The systems have become technically very good, perhaps with an overdominance of hospital care contributing to high costs (Kohn and White 1976). There are, admittedly, important differences in the evolution of the social policies in these countries (Esping-Andersen 1979) but they have, on the whole, produced good-quality health care at the lowest acceptable organizational level of the medical care system, although they have led to generally high levels of taxation in these countries.

Elsewhere in Europe, other systems of a mixed type have evolved (Babson 1972; Maynard 1975). In France, a fee-for-service system coupled with national insurance for reimbursement has developed: a product of an alliance between the individualism of *Médicine liberale* and *La Sécurité Sociale* (Webb 1982). Patients have, theoretically, free choice of doctor although, as in Britain, this is often governed by proximity and acceptance. Since 1973, there has been a socialist experiment to break the link in France between the provision of health care and money, based particularly on multidisciplinary medical teams working in health centres, an experimental one of which is at Grenoble. However, there have been various financial and political problems in developing 'free' medical care because a centre's income is still determined by the *number* of medical activities it carries out. Since the doctors in the Grenoble centre, for example, are attempting to *reduce* levels of investigation and promote preventive medicine, this means such centres will find it difficult to balance their budgets. Ironically, they could save the social security system money by avoiding excessive use of fee-for-service investigations, but there have, as yet, been few developments in what is essentially a 'conservative' French system.

Health care organization in the Third World

Sri Lanka

Sri Lanka provides an interesting Third World example of a social welfare state which, since independence in 1948, has had the commitment to provide free health care for all. It is also interesting that, to do this, it has pursued liberal policies with regard to the development of orthodox *and* traditional (Ayurveda) medicine. Outpatient treatment is available free at government hospitals and dispensaries and inpatient treatment is also available in non-paying wards. There has been a considerable drive to extend access to health services to all of Sri Lanka's 15 million inhabitants who live with an annual GNP of only $US230 per capita. It is estimated that the average distance travelled to Western-type hospitals is 4 km and, in the case of traditional, Ayurvedic medicine, only 1.5 km.

Health care is state-financed and administered via three levels – local (peripheral), regional and national – co-ordinated by the Ministry of Health, which formulates overall health policies. The country is divided into 19 health regions, each under a superintendent, and each region is further subdivided into health areas under a medical officer of health. At the lowest levels are the public health midwife areas. Local health personnel are responsible in their own districts for implementing the Department of Health's overall policy objectives.

As a developing country, Sri Lanka is still faced with numerous health problems and the curative components of the system face great strain. There is a shortage of health personnel and great crowding in urban hospitals, which attract patients who often by-pass smaller institutions. This is in spite of the relatively good doctor: population ratio (Table 2.3). Preventive medicine is regarded as very important but still has major tasks in overcoming preventable diseases, such as malaria, which has undergone considerable resurgence in the 1960s (Learmonth and Akhtar 1979). There is something of a vicious circle as preventable diseases throw burdens on the health service and further reduce the finances which may be devoted to preventive services. Indeed, a large proportion of morbidity and mortality can be attributed to inadequacies in prevention (World Health Organization 1980).

Table 2.3   *Health care professionals:population ratio in selected countries*

| | Population per | | | |
| | Physician | | Nursing person | |
	1960	1977	1960	1977
Sierra Leone	20,420	–	5,900	–
Sri Lanka	4,490	6,750	4,150	2,050
Kenya	10,690	11,630	2,230	1,090
Thailand	8,000	8,150	4,900	3,540
Uganda	14,060	27,600	9,420	4,300
Spain	820	560	1,290	900
New Zealand	690	740	–	200
United Kingdom	1,090	750	420	300
Australia	860	650	–	120
Canada	910	560	300	130
USA	750	570	340	150
Norway	850	540	330	100
Saudi Arabia	16,370	1,700	5,850	950
Kuwait	1,150	790	190	290
China	3,010	1,160	2,850	480
USSR	560	290	340	210

Note: WHO recommended 'target' is 1 doctor per 10,000 population

Source: World Bank (1981)

The health care delivery system has a hierarchical structure in which the curative components (distinguished in planning from preventive services) are organized in a series of institutions ranging from central dispensaries, through rural hospitals and maternity homes, district hospitals, base hospitals, and general (provincial) hospitals, to teaching (specialist) hospitals (Orubuloye and Oyeneye 1982). The distinction between curative and preventive services has, unfortunately, resulted in some duplication of functions and has to an extent lessened the co-ordination of the two sectors. However, many people rely on Ayurveda as an alternative to both 'Western' sectors and there is still a training programme for these traditional practitioners, who are in great demand. Private medicine does still exist in the government-dominated Sri Lanka system although the State monitors standards. The private hospitals are usually known as nursing homes and generally cater for the affluent groups whilst the rural poor in the remoter areas are those who typically still lack access to good facilities.

Since October 1978, a new Country Health Programming (CHP) scheme has been initiated, to identify and solve health problems through a multisectoral framework, involving health, economic, social and physical training. Its aims include the improvement of water supplies and sewage disposal, and the extension of immunization, health education, malarial control and family planning. Under CHP, all government departments whose activities touch on the health sector are expected to contribute and participate. This is an encouraging example of inter-sectoral co-operation in a developing nation and may provide a future model, especially for preventive services where much of the hope must lie for achieving 'health for all by the year 2000', the WHO target. CHP involves elaborate targets. Programmes and schemes have been inaugurated but it is still rather too early to judge what the long-term results will be.

There are, inevitably, constraints on achieving the aims of CHP and of upgrading health services generally, particularly relating to shortage of trained personnel, drugs and modern equipment (World Health Organization 1980). Funds are also generally inadequate to maintain the existing health care infrastructure, let alone to extend CHP programmes, as health has continually to compete with other sectors for scarce resources. A practical problem has been shortage of transport to extend projects into the field and to supervise field staff (Orubuloye and Oyeneye 1982). As a result, health problems continue in spite of worthy intentions, a common problem in many developing countries.

Virtually all Third World countries now recognize the problems of inadequate supply of services, especially to rural areas. Hardiman and Midgley (1982) point to the experiences of Sierra Leone, in which problems of access are quite severe, particularly access to hospitals, which tend to be urban-based while the bulk of the population is rural. As a result, they found that only 37 percent of rural inhabitants compared with 47 percent of urban dwellers in a survey were using government health clinics – 56 percent of villages relying on traditional medicine, particularly for minor complaints, whilst some households had travelled 100 miles to reach a government hospital for more serious problems.

This type of result emphasizes the diversity of experiences in the Third World; Sierra Leone providing a contrast with the relatively good access to health services in Sri Lanka. Indeed, about one-third of non-users of government health services claimed they could not afford the charges, as services were not free. The overwhelming problem was, however, distance, which deterred 60 percent of respondents from attending, and meant that many rural residents were effectively denied modern health care. In countries such as Sierra Leone, it is important to ask why health care has not adequately reached all parts of the country. Inadequacy of resources is only one aspect of the problem; another concerns their allocation and expenditure. The overprofessionalization of the delivery of health and social services has created a substantial bureaucracy, and salary and administrative costs consume large proportions of earmarked expenditure. In addition, administrative inefficiency, wastage and corruption have created problems in many developing countries. There is a distinct tendency to spend resources on capital-expensive 'prestige' imported items which can usually only serve the small proportion who live in close proximity to them, and shortage of transport usually prevents their spheres of influence being extended (Hardiman and Midgley 1982).

Similar problems exist in other very poor developing countries such as Ethiopia, Burma, Malawi, Tanzania, Uganda and Kampuchea. However, even in the slightly better-off countries such as Nigeria, Kenya, Guatemala, El Salvador and Ghana, conditions may, in reality, be little better than in the very poorest countries and, in many of them, unstable political and social conditions can often hinder development. For practical purposes, the attitude of government and people can be as important in determining the efficiency of health care policies as GNP and levels of income.

Kenya

As a developing nation with a considerable colonial history, Kenya has long been in contact with Western medicine, dating from the work of the Church Missionary Society in the early years of this century. This provided a valuable basis for the early medical services, which had an important role to play in controlling the major infectious diseases. However, research and real Western medical intervention were intermittent and influenced very much by attitudes of the colonial power, Britain, up to independence in 1963. The interesting historical development of medicine and health in Kenya between 1920 and 1970 is well covered by Beck (1981).

Since independence, the system has still been influenced very much by overseas-trained professionals although this reliance is now steadily decreasing. This is a common picture in many developing nations which are usually unable to provide sufficient training facilities or personnel for their own needs. In 1964, 67 percent of Kenya's medical practitioners were European and even in 1972, 63 percent of doctors were still non-citizens, as were 27.5 percent of registered nurses (Diesfeld and Hecklau 1978). It is also quite usual for many developing countries, especially the recently independent, to embody public health and the right of access to health care in their new constitutions as an obligation of the State, even if it is unlikely to be able realistically to fulfil it. This recognizes that the health of a population and its productivity are closely linked and that both must be improved if the nation is to grow in independence. In this, Kenya has been no exception and health strategies were incorporated into its first five-year plans.

Financial arrangements were that health care for outpatients and hospital treatment for those under 18 would be financed by the State. Until 1969, the running of basic health services was the responsibility of the local authorities but, due to considerable expansion of facilities from 1970, the central government and the Ministry of Health took over the running of basic health services (hospitals, clinics and dispensaries) apart from those in five large municipal councils (Figure 2.5). This provided opportunities for reorganization but also imposed considerable burdens on the administrative hierarchy (Diesfeld and Hecklau 1978).

Kenya does illustrate a number of the problems of providing health care in developing countries but also indicates some directions forward. It is a multitribal and

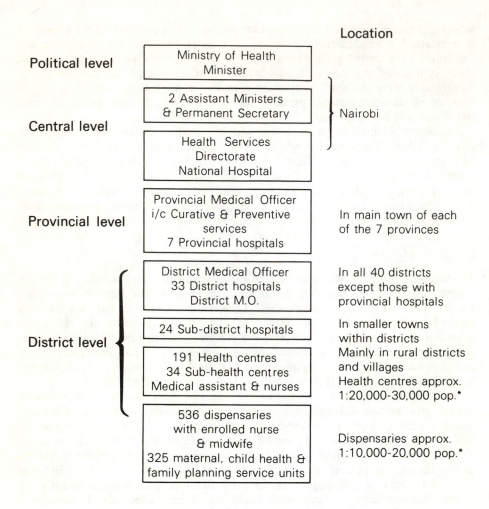

Location

Political level

Central level — Nairobi

Provincial level — In main town of each of the 7 provinces

District level

Box	Location
Ministry of Health Minister	
2 Assistant Ministers & Permanent Secretary	Nairobi
Health Services Directorate National Hospital	
Provincial Medical Officer i/c Curative & Preventive services 7 Provincial hospitals	In main town of each of the 7 provinces
District Medical Officer 33 District hospitals District M.O.	In all 40 districts except those with provincial hospitals
24 Sub-district hospitals	In smaller towns within districts
191 Health centres 34 Sub-health centres Medical assistant & nurses	Mainly in rural districts and villages Health centres approx. 1:20,000-30,000 pop.*
536 dispensaries with enrolled nurse & midwife 325 maternal, child health & family planning service units	Dispensaries approx. 1:10,000-20,000 pop.*

*There are considerable spatial variations in population:facility ratios

Figure 2.5   The health care hierarchy in Kenya

Source: after Diesfeld and Hecklau (1978)

multiracial society, which implies that different cultural groups' expectations must be met with regard to health services. It also has a very juvenile population: almost half of its 15.3 million population are under 14 years of age and crude birth rates in 1979 were over 50 per 1,000 (World Bank 1981). This naturally throws considerable strain on child health services, especially in rural areas, where almost nine-tenths of the population live.

As a result of population growth, there have been the usual environmental pressures on housing, sanitation and water supply in the major urban areas of Nairobi and Mombasa and consequent health problems. Primacy and the desire to procure modern training hospital facilities have, as in Sierra Leone, led to a concentration of health care expenditure in Nairobi in particular. The Nairobi Kenyatta National Hospital, for example, is the only national general and specialist hospital. In 1974, 52 percent of government-employed specialists and 50 percent of postinternship medical officers were working at this *one* hospital. This naturally creates something of an imbalance in access to specialist and higher-order care! Moreover, it may convincingly be argued that this is not the most appropriate level of care for many developing countries, which have needs which can be more readily met by less specialized dispersed care.

Eighty government hospitals provide only about half of hospital beds, the balance being made up of Church mission hospitals and some private hospitals. The overall bed:population ratio in the early 1970s was somewhat better than many developing countries had achieved, at about 1.25 per 1,000 persons, but the regional distribution of hospital beds was uneven. The Eastern and North Eastern Provinces, for example, had a relatively poor coverage because of their large areas.

Since the early 1970s, emphasis has been given to extending health services in rural areas, a policy which attained top priority in the 1974–1978 Health Development Plan. It is still the greatest challenge, however (World Health Organization 1980). Extension in the rural areas is mainly by means of health centres and dispensaries, some two-thirds being run by the Ministry of Health, and about one-quarter by the missions. At the end of 1977, there were some 191 health centres and 536 dispensaries, mainly in the rural areas, but the population:facility ratio fluctuates considerably amongst the seven provinces, from some 1:43,000 to 1:84,000. There is, in theory, a referral system through the hierarchy from dispensaries, health centres, subdistrict and district hospitals to the provincial general hospitals although how efficiently this works, even allowing for Kenya's relatively good transport network, is questionable.

As in many other developing countries, rural services extension has been attempted via the encouragement of self-help and community participation in health and family planning programmes. Additionally, paramedical staff such as nurses and midwives are used extensively to provide care. The provision of health centres, including mission hospitals with fewer than 20 beds, is planned at the rate of 1 per 25,000–50,000 inhabitants, although even this modest standard is rarely being achieved. In the less densely populated districts, in which health centres have to cover extensive catchment areas, travel and accessibility problems are considerable, and, in the west in particular, facilities sometimes fall qualitatively below standard.

Therefore, Kenya provides an example of the problems and some of the prospects for the poorer developing countries. Although, by contrast with many others, it has not chosen to support traditional medicine and encourages only qualified practitioners (Kikhela et al. 1981), the former is still undoubtedly relied upon by many tribal inhabitants, especially in rural areas. The nation has strenuously sought to modernize and, in recent years, to extend care to rural populations. Perhaps Kenya is fortunate, however, in having a relatively good base on which to build and a more stable form of government than many of its neighbours. Nevertheless, it is not immune from the shortages and increasing costs which have beset the health care programmes of many developing nations.

Thailand

Thailand provides an example of a developing country in which considerable strides have taken place in health care planning. Thailand has a population of some 46 million persons, A GNP of $US590 per capita, and a crude birth rate of 44 per 1,000. It is therefore in a slightly better financial position than many of the poorest countries although its demographic profile is much the same. Health services are provided by both private and public sectors in the ratio of 3:1 for capital expenditure. The private sector concentrates on curative and family planning services whilst the public sector's major concern is the provision of care in the 71 provinces in the provincial Health Care Services.

The provincial services have been designed to cater as efficiently as possible for a large rural population, using the relatively small number of trained staff available. To do this, the infrastructure of health services planning outside the capital, Bangkok, adheres closely to the administrative divisions of the country (Figure 2.6). There is a strong trend towards centralized control, the Ministry of Health being responsible for organization and administration of health services and, with Sri Lanka, Thailand provides an interesting example in the Third World of the aim of centralized control of health services with decentralized execution.

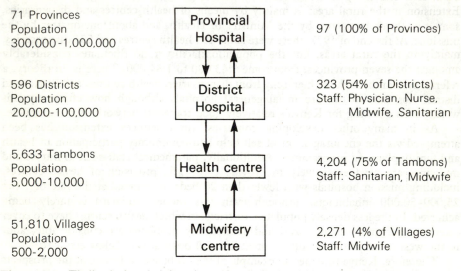

71 Provinces Population 300,000 - 1,000,000	**Provincial Hospital**	97 (100% of Provinces)
596 Districts Population 20,000 - 100,000	**District Hospital**	323 (54% of Districts) Staff: Physician, Nurse, Midwife, Sanitarian
5,633 Tambons Population 5,000 - 10,000	**Health centre**	4,204 (75% of Tambons) Staff: Sanitarian, Midwife
51,810 Villages Population 500 - 2,000	**Midwifery centre**	2,271 (4% of Villages) Staff: Midwife

Figure 2.6   Thailand: the administrative and organizational hierarchy of health care

In this hierarchy, not surprisingly, the basic village unit is the midwifery centre, given that 43 percent of the population is under 15 years of age. Each is staffed by an auxiliary midwife who provides general health care as well as mother-and-child services. These, supposedly, serve some 3,000 rural inhabitants at the village (and surrounding area) level. Health centres are established at the 'tambon' (subdistrict) level and each is manned by a male junior sanitarian plus a midwife, who provide minor medical care and important sanitary and public health information. The district level is a key tier, the first level to provide inpatient and outpatient hospital care by a more complete health staff, in hospitals of 10 beds or more. The provincial hospital lies at the apex of this hierarchy and provides, in the provincial cities, quite extensive medical and surgical care.

One of the major problems facing this apparently well-organized system has been a shortage of qualified manpower: for example, only 263 of the 323 district hospitals are staffed by physicians. As a result, there has been a general low level of utilization of government health services, perhaps caused by a lack of confidence. There is also the feature of an extensive and popular traditional medical sector, well integrated into the Thai culture. It is estimated that this services directly or indirectly about half of the population's symptoms (although lower proportions of actual illness), especially in rural areas in which modern health services are either not available, too distant, or too expensive (World Health Organization 1980; Porapakkham 1982). However, the Western training of Thai doctors has led to the dominance of Western-oriented medical care in the official system.

Thailand also experiences problems in covering rural areas in which 85 percent of the population live; and recent five-year plans have emphasized the improvement of health at the village ('muban') level. In particular, primary health care has been promoted but, in a recent study, 56 percent of urban versus only 23 percent of rural persons used hospitals or private clinics with trained doctors. The differential availability of such care is the single main reason for this difference although rural attitudes and reliance on traditional medicine as a first resort may also in part explain it (Porapakkham 1982).

Problems with health care research in developing nations

There are a number of problems associated with research into health care provision and administration in developing countries additional to those found in other nations. A first basic problem concerns demographic, manpower and fiscal data which are frequently out of date, unavailable or plainly inaccurate (Casley and Lury 1981). Financial data, in particular forward projections, are often inaccurate or wildly optimistic. Secondly, official statements are often out of step with the reality of any given situation, whether in terms of facility provision or in health levels. This is partly caused by a lack of data and partly by a natural desire by governments to portray their countries in a favourable light for some purposes, yet in a needy state for foreign aid. National and political pride has frequently to be balanced against the need to secure overseas assistance, particularly in very expensive health service provision. Therefore, at times the real situation may be worse than officially depicted, at other times better!

There are problems with pursuing basic utilization research in some countries, relating to language and cultural problems as well as to a shortage of trained assistance. As a result, we have found little evidence from Third World countries to include in Chapter 6 partly because of a paucity of reliable published research. In addition, much research, to be officially approved, needs to meet criteria prevailing at any given time. Research into family planning may, for example, be disguised as research into 'patterns of fertility'. This decade, the WHO objective is to provide adequate drinking water for all persons so research on this topic is current and generally acceptable. This is not to say that medical geographers should not follow current trends – indeed, these are frequently very practical and applied – yet some types of equally valuable basic research may not be encouraged simply as being unfashionable or politically unacceptable.

There is by no means a uniformity of experience in health care in developing nations. Some more recently developed countries, notably the wealthy oil-exporters of the Middle East, face problems not of health care costs but of how best to introduce to their people (particularly to their relatively fewer rural dwellers) highly technical health care facilities imported from the West. These problems are of a different nature and magnitude to those of, say, Thailand, the Philippines, Kenya and Uganda, in which there are urban health problems but also great problems to be overcome in extending basic health care to the rural bulk of the population. These and other problems should therefore be kept in mind when research is contemplated or when official statements are issued with regard to health care, particularly in developing countries but also in certain developed and socialist nations.

Socialist health systems

Many people would be sympathetic in principle to the socialist objective in health of providing a good standard of care for all. However, it is sometimes unwise to draw too sharp a distinction between 'medicine under capitalism' and 'medicine under socialism' as there are many shades of each (Navarro 1976; Parmelee et al. 1982). The examples here are drawn mainly from two different countries within the socialist world – the USSR and China – and they illustrate the common themes but also the variety existing even in 'medicine under socialism'.

The USSR

Health services have always been given a high priority in Soviet political and economic policies although it is difficult to compare with any meaning the pre-Revolution and post-Revolution health care of the Russian people because, since 1917, enormous strides have been made in medicine and social care internationally. It is possible, however, to identify a number of basic principles upon which medical care has been developed in the Soviet Union.

Health services are an integral part of the national socio-economic plan, and take their place amongst other national plans on which the Soviet State is organized (Fry 1971). Protection of the people's health is established as one of the most important tasks of the State in the current legal code (Ryan 1978). Medical care is virtually free at the time of use although nominal charges are made for drugs and appliances to some citizens. An important aim has been to extend medical care as far as possible given the physical size of the country, using various modes of transport and types of medical providers as necessary, to provide access for as many people as possible. Public participation is encouraged, to an extent, as is preventive medicine. Much care is, however, provided by specialists, and the USSR has, on paper, one of the best population:doctor ratios in the world (Table 2.3). Yet there are important reservations about the quality and nature of some care, particularly, for example, psychiatric medicine. It is claimed that health services account for 7.5 percent of the Soviet budget although this, of course, is difficult to verify. In line with the ideological basis of the Soviet State, there are no private or philanthropic health care facilities such as those that coexist with the public sector in the British NHS. In view of this, it might be inferred that a single, monolithic organization exists, capable of securing an optimal distribution of infrastructure, equipment and manpower. However, this requires qualification as a number of organizational frameworks seem to exist for health care delivery and there is evidence to suggest that the State would like even greater uniformity and central control to develop (Ryan 1978).

Nevertheless, to an outsider, the administration and organization of Soviet health services does appear to be quite uniform and to correspond with the general structure of central and local government (Fry 1971; Shannon and Dever 1974). This is to be seen in other centrally planned economies, such as Poland, although some, such as Yugoslavia, have gradually moved from a highly centralized system to one which gives more formal decision-making power to the republics and communes (Parmelee et al. 1982). In the USSR the spatial organization tends to follow the structure of central and local government, based on population size.

The State Planning Committee (Gosplan) provides central co-ordination and planning of health care, working closely with the Council of Ministers, of which the Minister of Health is a member. This level provides national policy and budgetary objectives, and the Ministry of Health has a large staff, many of whom are physicians. Below this is a very hierarchical structure with accountability within tiers and upwards (Figure 2.7). If ideology permitted greater decentralization of power, perhaps fewer health administration levels would be required. At present, below the central body are 15 Soviet Republics, each with a Ministry of Health. Within the Republics, detailed administration is achieved at a regional level, the basic unit being the 'oblast', or the 'kray' in the RSFSR (Dewdney 1979). Each contains between 1 and 5 million persons and has a regional health department and a chief medical officer who administers all the polyclinics, hospitals and public health services in the region. Below this level there are clinical department chiefs in each oblast, responsible for work norms and quality: performance criteria which are very heavily emphasized in the Soviet system.

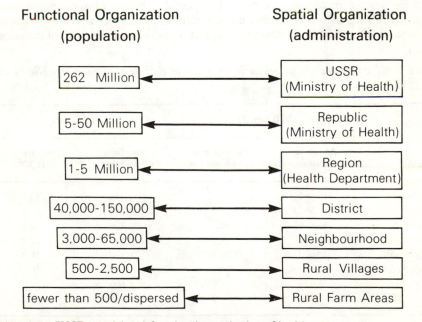

Figure 2.7 USSR: spatial and functional organization of health care
Source: after Shannon and Dever (1974)

Day-to-day management occurs at the district ('rayon') level, for populations of between 40,000 and 150,000. This level is important as it usually has a number of hospitals, one of which is the central district hospital. The district facilities are under the control of a chief district physician, again accountable upwards. At the neighbourhood ('uchastok') level, 3,000–65,000 persons are provided with polyclinics, functionally serving some 4,000 persons each on the average. There is no choice of physician and the polyclinic will be the first point of contact in most cases. However, access to care can be via separate polyclinic doctors for women and children, this specialization rather fragmenting primary care, particularly in the urban areas (Figure 2.4). Specialties within neighbourhoods are allocated on a population:doctor ratio, which again places heavy reliance on planning norms (Shannon and Dever 1974). A prominent feature of Soviet health protection is the provision of a large volume of medical care at the place of work. Many employing agencies make available the funds for construction of polyclinics and hospitals purely to cater for their workforces. This can, however, lead to unnecessary proliferation of small units. These will often eventually be staffed and maintained by the Ministry of Health although some remain the responsibilities of other government departments which, in 1970, accounted for some 4 percent of total hospital beds (Ryan 1978).

Urban residents generally have good access to a range of facilities, if little free choice. Unfortunately, as in many large countries, some rural districts have been difficult to supply with health care (Kamerling 1976; Ryan 1978). There are also considerable variations in standards amongst the 15 Republics, marked rural–urban disparities, and evidence that some of the ethnic areas of the Soviet Union, such as in Soviet Central Asia, are relatively poorly supplied with medical care, in spite of the massive increases in provision between 1950 and 1974 (Table 2.4). To overcome some of these problems, travelling teams of doctors periodically visit rural district hospitals. However, it is the use of 'feldshers', medically trained assistants who are not doctors, to extend health care in rural areas that is one of the best-known features of the system.

Table 2.4 *Examples of variations in numbers of doctors by Union republic in 1950 and 1974*

Republic	1950 (end)		1974 (end)	
	Absolute number	Doctors per 10,000	Absolute number	Doctors per 10,000
RSFSR	160,200	15.6	451,100	33.7
Ukrainian SSR	51,700	13.9	150,800	30.9
Uzbek SSR	6,600	10.2	33,400	24.4
Kazakh SSR	6,400	9.5	37,100	26.2
Georgian SSR	9,500	26.7	19,100	38.8
Latvian SSR	2,900	15.1	9,500	38.3
Tadzhik SSR	1,300	8.4	6,600	19.6
Estonian SSR	1,500	13.5	5,100	36.0

Source: after Ryan (1978)

These feldshers may be responsible for diagnosis and referral in primary care for small populations of perhaps 1,000 persons or fewer.

Nevertheless, there is little doubt that, in the USSR, as in China, access to, and quality of, health care is at its best in and near the urban areas. Good features of the Soviet system include the theoretical equal availability of care for all, the emphasis on planning, and the apparently very favourable ratio of hospital beds, facilities and personnel to population. The poorer features are that there is no free choice of physicians and considerable fragmentation of care by specialization, as over 90 percent of Soviet doctors are classed as 'specialists'. In addition, there is great reliance on stipulated norms and, presumably, relatively little work satisfaction for some doctors (there is, for example, considerable turnover amongst district physicians). There also remain many intractable problems in supplying rural areas with care similar to that of urban areas and, indeed, there is considerable scepticism as to the quality of much care administered and supposedly available. However, medical research has developed and there have been experiments with care delivery, such as 'hospital at home' treatment, although there is still generally lacking a doctor such as the British GP responsible for an overview of patient care.

The lasting impression is of an accordance of the Russian health service with the political organization of the country, as might be expected in the first communist economy. However, within this, some variations do occur and innovations develop and, in spite of persisting differences, there has been considerable convergence in standards of provision between the Republics during the Soviet era (Smith 1979). Indeed, Shannon and Dever (1974) suggest that the Soviet Union has emerged as a major world leader in providing its population with an equitable distribution of, and accessibility to, health services. Amongst recently emergent nations, the same can be argued for China, which is now discussed.

China

The Chinese revolution of 1949 brought an era of relative stability to a country that had been beset by internal strife and foreign incursions for over 100 years. The Chinese communist government immediately established a set of principles in the early 1950s which, to this day, form the ideological basis of health services development: medicine is to serve the working people; it must be integrated with popular mass movements; preventive medicine must be accorded priority; and, important to the current system, practitioners of traditional Chinese medicine should be 'united' with practitioners of Western-style medicine (Sidel and Sidel 1979).

There was much room for improvement after 1949, when China was still the 'sick man of Asia', riddled with many infectious diseases, periodic famine and malnutrition. In just over 30 years, China has moved from a society in which death, abject poverty and all forms of diseases were clearly visible, to one in which there is today very little evidence of gross ill-health and where there are few of the psychological malaises associated with developed countries. However, in the same short time the main causes of death have switched from pestilence and starvation to cancer, stroke and heart disease. This is a remarkable change if one considers that the population is over 1,000 million and very unevenly distributed in geographical, rural–urban, ethnic and linguistic terms.

The development of health services in China has not been uniform, however. In the early post-Revolution phase, modernization followed the Soviet model: the provision of high-technology, capital-intensive medical care, with Western-style training and specialization of doctors. By 1957, some 850 hospitals averaging 350 beds had been built – certainly a considerable achievement in scale and pace (Sidel and Sidel 1979). However, Sino–Soviet differences emerged and, with Mao's Hundred Flowers Campaign of 1956–1957, dissatisfaction with the direction of (Soviet) communism was expressed (Yong and Cotterell 1977). Health care, like all other sectors of Chinese society, came under close scrutiny. Medical technology was still poor when compared with most developed countries, but, at the same time as trying to establish a modern base with scientific medicine, China began attempting to integrate its ancient system of traditional medicine with this sector in a way few other countries have yet achieved. Another particularly Chinese approach was the organization of mass participation in campaigns to eliminate pests and disease. By the mid-1960s China had established the basis of a socially oriented, prevention-oriented health care system (Sidel and Sidel 1979).

The Cultural Revolution of the late-1960s criticized complacent administration in many sectors and the Ministry of Health did not escape. It is interesting that this turbulent re-evaluation occurred at the time when there was increasing international concern about the rising costs, inequitable distribution and dubious value of high-technology scientific medicine, and China's second post-Revolutionary health care direction commenced (Lucas 1980). This stressed the expansion of non-technological, low cost care, especially in the rural areas which has missed many of the improvements up to 1965. During the Cultural Revolution, the Chinese medical system underwent the disruption and self-criticism that touched almost every facet of life. It was disruptive to medical eduction and technical development although, by 1970, medical schools were readmitting students to be taught specially shortened medical courses to redress this.

In addition, the Cultural Revolution brought to the fore new forms of auxiliary medical workers, the primarily agricultural 'barefoot doctors' and health aides, production workers ('worker doctors') and housewives ('red medical workers'). The barefoot doctors are paramedics with a basic three- to six-month training for preventive, educational and minor curative work although, since the late 1970s, movements have been fostered to upgrade and improve their training and to standardize their work. Commune health facilities in rural areas are staffed by physicians and other personnel and generally have a range of specialist treatment available even if food for patients may need to be provided by their families!

In referrals, both rural and urban areas today have fairly clear and standardized procedures. There is a decentralized network of health centres at the production brigade level, staffed by barefoot doctors who may refer patients to hospital. In urban areas, the red medical workers have a similar role, especially concerning health education and birth control although their training is only about one month. Each urban neighbourhood has a hospital for major care but stress in both urban and rural areas is on decentralization of services and on collective responsibility for health. There is important interchange of health care personnel between locations and institutions, in which urban physicians work and teach in the countryside and paramedics come for training to cities (Parmelee et al. 1982). Today, there is increasing emphasis on appropriate and reliable scientific training and management rather than political indoctrination, part of China's modernization programme that spans not only health care but many other aspects of the economy (Phillips and Yeh 1983).

As might be expected, the Chinese State, often via the medium of collective enterprises, finances most of health care. There is responsibility on the individual and family in many cases to provide food and clothing whilst in hospital, and there is recent evidence that some forms of private medicine may reappear, although officially phased out in 1958 (Parmelee et al. 1982). The Chinese doctor has access to relatively few high-technology procedures and modern drugs yet he is today respected as a source of medical information. However, in spite of vigorous attempts to equalize access to care, the rural peasantry remain at a distinct disadvantage,and rural residents who live near to towns are the most likely to receive urban-based care (Table 2.5). In addition, rural residents are more likely to be treated with less expensive traditional medicine although this is admittedly very popular for certain conditions throughout Chinese society (Kleinman et al. 1975).

Table 2.5 *Estimated distribution of Chinese medical personnel in late 1970s*

Types of medical personnel	Estimated medical personnel/sector		Estimated population served/medical type	
	Urban Sector	Rural Sector	20% Urban (190,000,000)	80% Rural (760,000,000)
1	350,000	15,000	543	50,667
2	700,000	85,000	271	8,941
3	NA	1,800,000	NA	422
4	NA	4,000,000	NA	190
Mobile	917,000	183,000	207	4,153
Traditional	66,000	262,000	3,167	2,901
Total personnel	2,033,000	6,345,000	–	–
Population per medical personnel	–	–	94	120

Types of medical personnel:
1. Higher medical graduates
2. Intermediate medical graduates
3. Great Leap Forward and Cultural Revolution peasant doctors and barefoot doctors
4. Spare-time public health workers and rural midwives

Source: Lucas (1980)

China's health services are, therefore, working on a number of worthy principles (Sidel and Sidel 1979). These have all been underlaid by changes in control of resources and altered priorities in Chinese society since the Revolution, and by improvements in social, economic and nutritional standards nationwide. The overriding principle has been to extend medical care to all, by use of mobile medical teams, barefoot doctors and interchange of personnel between town and country.

Allied with this is decentralization of services to the lowest levels at which they can be provided and establishing clear referral channels to upper levels of the health care hierarchy. 'Deprofessionalization' and 'mobilizing the masses' are also central themes in organization of current Chinese society and health care although there is some recent evidence of a realization of the importance of co-ordinated research in medical care in particular. Finally, China is a world leader in integrating aspects of traditional medicine such as herbalism and acupuncture with its modern health sector. Therefore, 'continuity with the past' has been important but even more crucial is the fact that China has experience of the means of conjoining traditional and modern medicine which could prove invaluable for many developing countries where such systems do not always provide mutual support, as discussed in the following section.

Traditional or indigenous medicine

Most Westerners tend to regard medical care as being provided exclusively by personnel who have a 'scientific' background with training in medicine, surgery, anatomy and pharmacology. Health care is mainly seen as 'ill-health' treatment, to be issued and obtained as almost any other goods or service. However, for a very substantial proportion of the world's population, this type of scientific treatment is but one aspect of a more holistic approach to health care as part of life and customs. This occurs mainly but not exclusively in developing countries, and a number of somewhat vague terms have emerged to describe the provision of health care from what generally have been viewed as 'unscientific' sources. The terms 'traditional medicine' and 'indigenous medicine' have been used to indicate ancient, culture-bound health care practices which predate the application of scientific medicine or allopathy. Other frequently used synonyms are ethnomedicine, fringe, folk, alternative, unorthodox, pre-scientific, or tribal medicine. Many countries regard the practice of such medicine as 'unofficial' and, sometimes, as illegal. However, none of the terms above is entirely satisfactory because each implies that practitioners will refer to a common body of knowledge, will have training, and will be of a more or less uniform approach (Maclean and Bannerman 1982). Nothing could be further from the truth as the range of sophistication and religious or cultural commitment found in various types of 'traditional' medicine (for want of a better term) is enormous. There are, for example, important differences between, on the one hand, complex systems which require a specific attitude to life and health and are strongly culturally determined – the traditional Chinese systems, say, or some South Asian systems such as Ayurveda (Crozier 1968; Jaggi 1976; Lee 1981, 1982; Jeffery 1982) – and simple home cures or 'quack' remedies on the other hand.

Although for long a subject of social anthropological interest (Mair 1965), concern about the place and utilization of traditional healers in nascent health care systems today has recently been growing rapidly amongst a wider group of academics and health care planners. This is particularly so since it is now realized that few developing countries can afford the spiralling costs of providing modern health care for all their populations and that adequate resources in terms of (scientifically) trained manpower will hardly be available before the end of this century. The target of 'Health for all by the year 2000' is therefore unlikely to be met from modern medicine alone (World Health Organization 1981). In addition, since there are sizeable ethnic minorities who have 'imported' traditional ideas to developed countries, knowledge of these systems is increasingly relevant. It will also be seen in Chapter 6 that the use of 'folk' remedies can characterize certain groups and can be an important stage in the receipt of care. Therefore, it can be safely assumed that indigenous medical beliefs and practices have been *and are* an integral part of many human cultures and display a range of sophistication from the quasi-scientific and truly curative to those dependent on faith healing, or upon supernatural or spiritual healers (Ramesh and Hyma 1981; Stock 1981).

What are the types of traditional medicine?

In some ways, the use of the word 'medicine' in this context is misleading because, in Africa, for example, many practitioners are deeply involved in the maintenance of social order and in preserving cultural institutions (Kikhela et al. 1981; Maclean and Bannerman 1982), whilst traditional 'medicine' in some Indian and Chinese systems springs out of attitudes to life and religious cultures (Kleinman et al. 1975; Ramesh and Hyma 1981). As such, healers have a more important cultural role in the community than most Western-style physicians. However, the aim in this chapter is to introduce some of the main types of traditional medicine. Their roles in social control are discussed by Eyles and Woods (1983).

It is perhaps helpful to consider that there are four general categories of indigenous practitioners (Neumann and Lauro 1982): 1. Spiritual or magico-religious healers; 2. Herbalists; 3. Technical specialists such as bonesetters; 4. Traditional birth attendants (TBAs). Although the functions of some of these practitioners might certainly overlap in any given circumstances and certain types of healers might not fit under any heading, this is a useful classification to bear in mind during the following discussion.

It should be realized that unconventional or unorthodox healing practices can be adopted at certain times by individuals in all types of countries – within developed countries there are considerable followings of faith healers and fringe medicine. More tangible types of unorthodox medicine exist, too, some of which are increasingly recognized as having curative value. Chiropractors, for example, can cure various types of musculo-skeletal disorders by manipulation of the spinal column, and osteopathy is, in fact, allied quite closely to some aspects of physical medicine. Osteopathy is increasingly popular amongst Westerners suffering from back pain for which conventional medicine frequently offers little relief. Homoeopathy – the treatment of diseases by small quantities of drugs that usually produce symptoms similar to those being treated – is also recognized as a type of herbalism which can be valuable in certain conditions. Finally, herbalism as a more general subject underlies much modern pharmacology and chemotherapy. It is interesting to note that naturally occurring herbal drugs often produce fewer side-effects and medicine-created (iatrogenic) diseases than do modern synthetically derived drugs with similar curative properties.

In the West in particular, the proportion of the population relying totally on unorthodox treatment is always relatively small. In developing countries, however, this need not be so and, particularly in rural or tribal areas, non-scientifically trained healers may provide almost all of the available care. It has even been suggested that in India, for example, the masses were denied access to Western medicine almost as a means of colonial control (Banerji 1979). Whatever the validity of this argument, it has been estimated that in contemporary India organized health services provide only 10 percent of medical care and a further 10 percent is provided by qualified physicians in towns and cities. The balance is split between home care and indigenous medical practitioners (Taylor 1976; Ramesh and Hyma 1981).

In the Indian subcontinent, the range of indigenous medicine is considerable, and systems have evolved and progressed even in the urban locales. There are two parallel systems: the modern, usually referred to as allopathy and homoeopathy; and the indigenous, comprising Ayurvedic, Siddha and Unani systems. Naturopathy and Yoga also have their followers. Ayurveda is used all over India, being a traditional Hindu system, whilst Siddha is more commonly used in the south in Tamil Nadu and neighbouring states. The Unani system is found predominantly in areas of Muslim culture (Ramesh and Hyma 1981).

Ayurveda utilizes herbs, minerals and diet restrictions in the treatment of illness and, with the somewhat similar Siddha system, is practised by indigenous medical practitioners (IMPs) known as 'vaids'. The Unani system, known as the Greek/Arab system, was brought into India by the Muslims and also uses herbs, minerals and metals in its treatment and is practised by 'hakims'. It is based on the ancient Greek theories of the body's humours and has links with old-established concepts of Hippocrates and Galen (Jaggi 1976). This century, official attitudes to traditional medicine have varied but many aspects are still taught in colleges and a number of certificates and degrees are available in Indian medicine, plus registration for those practitioners who have served apprenticeships and worked under guidance for certain set periods. There are a number of legal bases for the practice of traditional medicine in many countries and India provides one good example. India has, on the whole, a good supply of indigenous medical practitioners, and some estimates are given in Table 2.6, the total being slightly fewer than another estimate of 500,000 practitioners (World Health Organization 1978b. It is estimated that some 7,000–8,000 qualified practitioners of Ayurveda, Siddha, Unani and Homoeopathy enter their professions each year (Ramesh and Hyma 1981).

Table 2.6 *Registered practitioners in Indian systems of medicine and homoeopathy, 1977*

	Institutionally qualified	Not Institutionally qualified	Enlisted
Ayurveda	117,765	105,344	–
Unani	10,262	20,138	–
Siddha	1,559	16,569	–
Homoeopathy	19,871	74,166	51,397
Totals	149,457	216,217	51,397

Grand total: 417,071

Source: after Jeffery (1982)

The relative importance of these types of traditional medicine varies regionally in India as well as in urban and rural localities, and there are few easily available geographical studies. However, a 10 percent study of the 956 traceable registered

IMPs in Madras in 1976 showed that 36.4 percent followed the Siddha system, 33.2 percent Ayurveda, 10.4 percent Unani, and 20 percent 'integrated' systems. Most of the practitioners were aged between 40 and 60, and most lived close to their served population; few had a telephone, car or large home. The majority ran simple surgeries in their homes and the main locations were in the older residential quarters. The upper- and high-class neighbourhoods did not, by and large, attract the IMPs. Their consultations were much more leisurely than those of modern practitioners, and their attitudes to patients and their spatial availability (near to the needier families) more acceptable. In addition, as is frequently the case in traditional medicine, they charged less for their services than modern physicians and they provided many dietary prescriptions, which are often expected by people of Indian culture when they are ill. However, their registration is far from complete and, so far, only lip-service has been paid to promoting their integration in the modern system (Ramesh and Hyma 1981).

Other important practitioners, 'traditional birth attendants' (TBAs), are ubiquitous in developing countries and are taken for granted as midwives. They are probably more experienced in child deliveries than many Western-trained personnel. In the Philippines, for example, over 75 percent of all births take place under the supervision of TBAs (locally called 'hilots') and the government simply has no alternative to recognizing their existence and, sensibly, involving them in other primary health care activities such as notifying communicable diseases, organizing mothers' classes, assisting in immunization and registering births (Vuori 1982b). In at least 44 countries, including India, Malaysia, Pakistan, Bolivia, Brazil, Sierra Leone, Nigeria, Panama, Paraguay and Peru, traditional midwives have been given formal training to upgrade their performance although examples of their incorporation into modern-sector projects and programmes are rather fewer (Pillsbury 1982). Utilizing the existing and trusted resources provided by TBAs is evidently a sensible way to increase the effectiveness of primary health care. However, uncritical optimism about the use of TBAs should be guarded against as there have been noted problems of supervision and education: for example, in implementing a programme using 'parteras' in Nicaragua (Heiby 1982).

Within the Chinese culture, as in the Indian, a number of important traditional medicine systems exist (Crozier 1968; Kleinman et al. 1975). One of the best known and least understood is acupuncture: the treatment of ailments and pain relief by sticking needles in strategic sites of the body. This is a very ancient art in East Asia, and is increasingly realized to have some value in curative terms. Classical Chinese medicine is based on cosmological ideologies, formulated on a large body of principles, scientific and unscientific. In the past, apprenticeships included studying literature and acquiring clinical experience but, since there existed only a relatively small number of trained 'medical officers' who had passed a State examination, their impact was limited to major provincial cities. Up to the middle of the twentieth century, most urban and rural poor had to rely on simple empirical or magico-religious remedies (Lee 1982).

Confucian scholars would also practise curative methods out of humanitarian motives and they commanded high prestige in pre-Communist China. A third category of healer was the private practitioner who worked in local communities and who had rather less training than the official medical officers but was also mainly urban-based. Today, in Chinese societies, many herbalists and traditional practitioners still exist (Lee 1975; Topley 1975; Phillips 1983a, 1983b). Some work mainly as bonesetters and are experts in manipulative 'technical' medicine, often being learned teachers of the Chinese martial arts.

In the modern-day Chinese People's Republic, traditional healers are still used and, as noted earlier, there has been particular emphasis on the incorporation of appropriate aspects of Chinese medicine into China's modernization drive. Mao Tse–tung, in particular, held very strongly to this belief, since a greater number of people could be served by this rather lower level of care than could be by high-technology care based on the Soviet model (Sidel and Sidel 1979). However, the barefoot doctors, for which modern Chinese health care is perhaps best known, have mainly be drawn from the ranks of the more literate peasant workers. Contrary to common belief, very few were originally herbalists but were people who tended to hold idealogical commitment and belief in community or mass participation (Wilenski 1976; Pillsbury 1982). A minority of barefoot doctors had experience as herbalists but all have been encouraged to use herbal as well as Western medicines in their activities.

Attitudes to traditional medicine

In some developing countries, perhaps too great an emphasis has been placed on expenditure to secure capital-intensive 'scientific' health care which will never reach the mass of the people. In the past 25 years, medical care in developed countries has been subject to enormous cost increases of between 300 and 500 percent (see Chapter 8). Cost increases have been even higher in many developing countries. As a result, in some countries, 80 percent or more of health budgets are being used solely to support major hospitals whilst the bulk of people are very inadequately covered by overstretched clinics (Elling 1981).

A concentration on primary care is probably more effective in terms of collective outcome in these countries and it is at this level that co-ordination and complementary developments between traditional and modern systems may develop. This does not preclude some traditional practitioners from being specialists in some types of care (particularly of mental disorders), but most appear to work best at the community level (Kikhela et al. 1981). A problem is that many developing countries feel that traditional medicine is identified with socio-economic underdevelopment and indicates backwardness (Vuori 1982b). Official health services and professionals have often disparaged traditional medicine, fostering an atmosphere amongst the ruling élites in many countries that does not favour its furtherance. Nevertheless, rather than being 'dead as the dodo', 'traditional medicine seems to be alive and thriving – to such an extent, indeed, that it has been estimated to be the principal, if not the only, source of medical care for two-thirds of the world's population' (Vuori 1982b, p. 129). Some authors even go further and suggest that 'pre-scientific' medicine may treat up to 80 percent of the world's people (Scarpa 1981).

As a result, the WHO is now paying considerable attention to its medical efficacy: a much more problematic area than its socio-cultural value. The WHO has now firmly espoused the potential of traditional medicine, subject to rigorous scientific investigation. In 1979 it set up the guidelines for the development of a programme to promote traditional medicine, having been interested in this since the 1950s, when it became involved in training 'hilots' in the Philippines as part of the country's midwifery programme. The programme aims to foster a realistic approach to traditional medicine in order to promote its contribution to health care and to explore its merits scientifically. This should enable the encouragement of its beneficial aspects and discouragement of harmful effects. The programme's most important aim, however, seems to be that of promoting the integration of proven, valuable knowledge and skills of traditional *and* Western medicine (Vuori 1982b).

This seems essential if steps are to be taken to achieve 'health for all' this century, but there are numerous pitfalls and problems in this integration. In India, for example, foci of concern have been whether to incorporate indigenous practitioners such as vaids and hakims into the State medical service or whether to train a separate cadre of community health workers, *and* how to register existing practitioners and prevent unregistered practice. In addition, it is questioned whether indigenous colleges should include Western scientific training, and vice versa (Jeffery 1982). If this does not happen, there is little chance of future mutual respect between the two parts of the medical profession, 'professional recognition' of traditional healers being a major stumbling block in many systems (Lee 1975; Topley 1975). There is evidence, however, that in socialist-oriented developing countries such as China, a lack of resources coupled with a desire to build for the whole country a good health base can bring about an effective merger of modern and traditional systems, with an emphasis on making the merged system serve the people. On the other hand, in socialist-oriented systems at higher resource levels, such as Russia and Cuba, there is an emphasis on predominantly modern, technical medicine, even if the form of control and delivery differs from that of capitalist countries (Elling 1981). Therefore, as stressed earlier in the chapter, the political economy of nations and their cultural hegemony can very much influence the mixes of traditional and modern medicine, which Elling (1981) has organized into a simplified chart (Table 2.7).

Table 2.7 *Political economy and 'mixes' of traditional and modern medicine*

Political economy	Resources	
	Low	High
Socialist-oriented	Merger of **T** and **M** with system made generally available, e.g. China	**M** predominates with system made generally available, e.g. Cuba, Poland, USSR
Capitalist-oriented	**M** for ruling class; **T** of poor quality for peasant and working classes, e.g. India, Mexico, Philippines, Thailand, Zaire, etc.	**M** predominates but is unequally distributed in favour of ruling class. What **T** there is, is more employed by working class, e.g. USA, UK. Both **M** and **T** for working class of poorer quality

T = traditional medicine   **M** = modern medicine

Source: after Elling (1981)

Vuori (1982b) provides a very useful regional summary of the major systems of indigenous medicine which seem to be thriving in Africa, Central and Latin America, the Western Pacific and South-East Asia. In Europe, there is little national support for the use of traditional medicine other than osteopathy and chiropractic, and a similar picture exists in North America. The eastern Mediterranean region of the WHO provides an interesting bridge since there is a rich cultural heritage in traditional medicine, especially in rural areas, but official attitudes vary greatly from recognition and support to rejection and even suppression. However, a number of countries have developed training programmes for TBAs and this branch of traditional medicine is, again, seen to be important in the mother-and-child health services.

Is it possible to generalize about the options available for health planners with regard to traditional medical systems? Kikhela et al. (1981) have attempted to analyse the attitudes taken in various African countries, and their classification outlines the potential range of attitudes. In reality, they view the two systems, indigenous and modern, as being complementary in all countries from the point of view of the user. However, rarely, if ever, does this seem to develop into effective official co-operation and integration from the planning perspective. This they hope will happen gradually but four main options seem to face policy formulators:

1. *Making traditional medicine illegal.* Some countries such as the Ivory Coast have opted for this approach which really implies a state of self-deception since recourse to traditional healers remains part of the people's basic medical care.

2. *Informal recognition* in which laissez-faire legal aspects are applied. Unfortunately, this involves ignoring an activity basic to citizens. Under this option, the State only becomes concerned in traditional medicine when it regulates its failings and abuses in the courts, so the healer, whilst practising therapy, has no practical protection. Having no legal status, traditional healers are generally contributing little to official health information and data. Countries such as Hong Kong and Singapore depend at least in part on traditional medicine, yet their legislation is generally merely restrictive: for example, attempting to prevent herbalists advertising themselves as 'doctors'.

3. *Simple legislation* may be introduced to govern the practice of traditional healers. Licensing announces who the healers are, which the population probably knows already, but it does not help the integration of the traditional and modern systems. It may enable research into the type and geographical distribution of healers but this option may do little to control the quality of traditional medicine. Unless serious attempts are made to control unlicensed practitioners, which can be costly, this option has relatively little value. Some countries, such as Nigeria and Ghana, have opted for individual licensing but without first modifying their health laws, which indicates that they do not intend to modify substantially their existing system.

4. *Gradual co-operation with healers.* This is the favoured and sensible option, largely supported by the WHO, which itself has a Unit of Traditional Medicine (Akerele 1983). Basic research into the nature, specialisms, extent and distribution of traditional practitioners must be undertaken, and the structure of traditional medicine must be established to permit integration. This process will be gradual because professional attitudes, particularly on the part of 'scientific' doctors, will only admit relatively slowly the proven aspects of traditional medicine. It seems that some traditional healers, because they are accepted and trusted, especially in remote and backward areas, may be able to introduce selectively modern medical treatments on behalf of the scientific sector. This has occurred in Nepal and other nearby countries. In addition, a number of countries such as Zaire, Guinea and Mali have begun to integrate their

modern and traditional systems at the lowest levels. Mali has established a National Bureau of Traditional Medicine within its Ministry of Health and, for more than 10 years, Senegal has been conducting serious research into the contributions of traditional healers to the care of mentally ill persons. TBAs are also important in Senegal (Sène 1983) whilst, in Mali, traditional psychiatry is increasingly seen to be valuable (Coppo 1983).

Within this fourth option, it is important that certain principles guide the gradual integration of traditional medicine into public health systems. In particular, health care will require administrative and spatial decentralization, with emphasis on preventive care. Open-mindedness on the part of both major partners in the new structure will be essential if it is to succeed and gradual integration should occur in all areas, urban and rural, where the two systems exist. If this does not come about, traditional medicine could retain the appearance of a rural peasant system, and take on an undesirable 'stop-gap' role. A final principle is to allow healers with specialized skills to act as specialists although, generally, most will act as the primary care, front-line health agents (Kikhela et al. 1981).

In the long run, it seems that many countries will, with international support, recognize the value and importance of integrating of biomedicine and ethnomedicine. This is a long way off in the main but healers can be very valuable in their milieus: they provide sensitive manpower and they are generally using cost-effective, appropriate levels of technology. This is not to say that all traditional medicine is useful but, if the effective parts can be distinguished, then it would be folly to ignore them. Research into international changes over the next few decades will make fascinating and very worthwhile geomedical study and it is important that it be developed soon.

3   Access to Health Care

Equity considerations in health care delivery

Interest in differences in 'level of living', in 'social well-being', and in related themes, has increased in geography in the last decade or so and has manifested itself in a number of ways. For most geographers, the critical thread running through this work is the emphasis, sometimes implicit but most often explicit, on areal variation in welfare. Therefore, although one might distinguish between work tied to a Marxist stratification of society (e.g. Harvey 1973) and that based on different, or perhaps less formally defined, models of society (e.g. Coates et al. 1977), the spatial theme is as a rule clearly evident in works of both types. Indeed, it would be naïve and unrealistic to attempt a separation between the spatial dimension and the other elements of societal organization. The role of geographical location in the determination of well-being should always be set within the broad context of the social, economic, and political forces operating in society (Dicken and Lloyd 1981). In the case of health care, Chapter 2 considered the emergence of different types of medical systems in the light of a range of factors, perhaps the most important of which is political economy.

The choice of dimensions or domains for inclusion in geographical studies of well-being depends upon a number of factors, including the philosophical stance of the researcher, the nature of the society under study, and the goals and objectives of the exercise, but in virtually all studies one or more dimension or domain relates to the provision of public services. Many social geography texts now indicate the range of indices devised to measure or define levels of living or quality of life (Coates et al. 1977; Herbert and Thomas 1982). This reflects a more or less worldwide phenomenon of the last quarter-century – the rise of the welfare state, the components of which were identified in Chapter 2. The citizens of various countries have as a result come to rely increasingly upon collective provision via social services for the satisfaction of their needs (Smith 1979; Dicken and Lloyd 1981). Among these social services, health care stands pre-eminent.

Smith puts the case succinctly:

> Health care is perhaps the most 'basic' of all services, for on this may depend whether a newly-born child lives or dies, whether we survive illness or accident and, if we recover, whether we retain full use of essential faculties or suffer permanent handicap.
> (Smith 1979, p. 246)

Even if incontrovertible proof is not available concerning the positive causal relationship between the supply of health care and the level of health of a community (Howe and Phillips 1983), there is strong evidence that the general public believes such a connection to exist (Fein 1972). There is often an implicit or explicit assumption in many health plans and policies that more health care will result in better health. Such belief is, in itself, sufficient to guarantee a special position for health care amongst the expanding complex of social services and endows a similar priority upon related questions of equity in delivery.

The appropriateness of equity as a criterion for the delivery of health care is well founded in liberal and social democratic philosophy, but what does equity entail? All too often, equity is taken to be synonymous with equality. As discussed further in

Chapter 8, these two concepts are not necessarily identical (Smith 1979). It is also unclear whether equality is to be in terms of (a) health care expenditures available for individuals, (b) the value of services actually used by individuals or (c) individual health outcomes (Fein 1972). The availability of services does not guarantee their use; nor does the use of equal amounts of health care guarantee a population of uniformly healthy individuals, so a 'pure' equity approach would demand a focus on health outcomes. However, the overwhelming emphasis to date has been upon questions of health service availability and/or utilization and rarely, if ever, directly upon the quality of outcomes associated with utilization. Studies of mortality patterns, especially of infant mortality, have touched upon the latter, of course, but it has consistently proved difficult to isolate the role of service quality within the complex of social, psychological and physiological factors affecting quality of outcome (Gross 1972). In many ways, the focus on availability and utilization has been by default, stemming from the inability to untangle the impacts of the various determinants of health status.

Forms of accessibility

Health care, like many public services, is not equally available to all individuals. This is because it is not a 'pure' public good (Cox and Reynolds 1974). One of the major reasons for public services being impure is geographical in nature. Demand for public services emanates from individuals, who, in aggregate, are continuously (though unevenly) dispersed across space, while most public services are distributed from discrete facilities with fixed locations (Dear 1974; Massam 1975). It follows that, *other things being equal*, uniform availability would exist only if every individual had immediate and uninterrupted potential access to facilities supplying needed services. Unfortunately, most people's backyards are not large enough to accommodate even a modest hospital, fire station, or library, and instantaneous and costless efficient transport exists only in the minds of science fiction writers and other savants. Moreover, 'other things' are rarely equal; geographical factors are not the only ones relevant to a consideration of health care delivery. Indeed, the interplay and relative importance of various sets of factors affecting access to health care is, in itself, an important area for discussion.

Supply of service is a prerequisite for accessibility. Unless services are 'available', there can be no consideration of the factors, geographical or otherwise, that differentially influence the access of individuals or groups to needed services. Availability in this most general of interpretations depends largely upon the state of medical technology and the allocation of resources to health care. The potential for meeting needs will exist only if medical science has the ability to ameliorate, control, or cure particular illnesses *and* if society is willing collectively to bear the cost of meeting particular needs. A decision on the latter may emanate from the market place, as it often does in the United States and currently in Australia, for instance; or from government, as it mainly does in countries like Canada, Great Britain, New Zealand, and the USSR. The various elements of the decision to supply public services have received substantial attention in the literature: representative works include Hitzhusen and Napier (1978) on the political processes and pressures, Merget (1980) on the economic influences, and Mechanic (1970) on the organizational and professional considerations. We shall consider the organizational and locational dynamics of health care supply later in this chapter and in more detail in Chapter 4 but, for the moment, we shall concentrate on the consequences, or more correctly, the modified consequences, of decisions affecting the supply of health care.

Donabedian (1973), among others, has recognized two main groups of factors affecting access to available stocks of health care – socio-organizational and geographic, as noted in Chapter 1. The former can be taken to embrace all attributes of the service such as cost, intake policy, specialization of the provider, and so on, that could give rise to differential access to health care on the part of individuals or groups. Among these socio-organizational factors, barriers based upon cost of care have often been considered to be of primary importance although this generally holds true only in free market systems (Fein 1972; Gross 1972; McKinlay 1972). This distinction has also been made in the geographical literature. *Locational* accessibility represents physical proximity and may be crudely expressed in mileage terms. *Effective* accessibility concerns whether a facility is always available or open, whether it is socially or financially available to people, and whether a person's time-space budget permits him to use the service (Ambrose 1977; Moseley 1979; Phillips and Williams 1984).

The importance of the economic barrier depends largely upon the nature of the health care system. In private enterprise systems, the economic factor is clearly of great importance as the financial resources of individuals and communities largely determine the nature and quality of care which they can obtain (De Vise 1973). However, even in countries such as New Zealand and Great Britain, which have a long and cherished tradition of 'free' health care, there are invariably barriers to accessibility based upon economic considerations (see Chapter 2). Consider, for instance, the implications for access of one- or two-year waiting lists for elective surgery within many (British) National Health Service regions in contrast to the one- to two-week wait for similar treatment within the British private system. Therefore, it would be hasty to think of cost barriers as being a purely American phenomenon, particularly given the strong association between income and other, social characteristics of individuals and groups, such as ethnic status. The ubiquity and strength of such associations across space and time provide grist for the Marxist mill!

The recognition of associations between economic and social factors affecting access to health care leads to obvious questions. How do they interrelate? What determines their relative importance? Do they change or can they be altered over time by health policy?

For a general discussion of the links between social and economic systems within a welfare framework, readers are referred to Harvey (1973), Smith (1977, 1979), and Knox (1982a). More specifically, the complex interrelationships between economic and social constraints on access to health care have made themselves evident in delivery systems that have moved from predominantly free-market to predominantly public financing. This occurred in Great Britain in 1948, for example, and in the various provinces of Canada beginning in the late 1950s. In connection with the Canadian experience, Enterline et al. (1973) reported a substantial increase in the use of health care by lower-income households in the Province of Quebec following the initiation of 'Medicare'. However, in a similar study conducted in the Province of Saskatchewan, Beck (1973) noted that, although the introduction of Medicare had apparently increased the *absolute* accessibility of lower socio-economic groups, there still existed a substantial disparity amongst the various strata of Saskatchewan society. This underlines the danger of an implicit assumption that all too often underlies discussion of the merits of 'free' medical care – that elimination of financial barriers will result in equitable access (at least, as measured by consultation rates) across social groups (Beck 1973). Indeed, the interface between economic and social barriers to utilization of needed services constitutes an important research focus in medical sociology (Cockerham 1978; Fiedler 1981) and will be considered in some detail in Chapter 6. At this juncture we focus on the interplay of these socio-economic factors (whatever their interconnection) with those of a geographical nature.

If health care was a simple good, that is, a uniform service was available at identical facilities, then the situation would be relatively simple, at least in principle. The absolute extent of socio-economic barriers, based upon cost of service or whatever, would depend upon prevailing societal attitudes, preferences, and priorities as manifested in the health care delivery system. In turn, the absolute importance of geographical factors affecting accessibility would depend upon the geographic separation of facilities and potential consumers, and the consumers' mobility. The relative importance of each set of factors would remain fixed as long as the absolute nature of all factors remained constant. Therefore, geographic constraints on accessibility would become relatively less important if socio-economic barriers were to become more pronounced, if the density of the service network was to be increased, or if the mobility of potential consumers was to be enhanced. Various combinations of the above are possible of course.

Health care is not a simple good, however, and its complexity is increasing steadily. As a result, rather than needing an amorphous and generalized service called 'health care', potential consumers may desire access to specific types of treatment or combinations of treatments, the availability of which depends largely upon the organizational attributes of the delivery system (Donabedian 1973). A key organizational attribute of the health care delivery system is its hierarchical structure and, almost without exception, the different systems outlined in the previous chapter depend on various forms of hierarchical referral or links between different levels of service.

The health care delivery hierarchy

Geographers often face the challenge of identifying the components or levels of hierarchies, most frequently in connection with urban systems and the ranking of urban places. The challenge is demanding because, although there may be a strong theoretical basis for the existence of distinct levels within a hierarchy, in practice, the boundaries between them often tend to be blurred and, indeed, to overlap. Additionally, there is invariably a possibility for more than one taxonomy; for example, the same urban system may reasonably be divided into three, five, or seven levels depending upon the classification or grouping criteria adopted by the analyst. Similar problems face those who wish to characterize the health care delivery hierarchy. For example, in certain systems, some clinics and doctors provide minor surgery and hospital-type procedures whilst, in others, hospitals become major sources of primary care, particularly through their emergency rooms.

As a preliminary to any discussion of the nature and merits of specific conceptualizations of the health care delivery hierarchy, it must be recognized that no single hierarchy exists across space or time. The great variety of philosophies and organizational structures observed among contemporary health care delivery systems provides ample evidence for cross-cultural differences. In spite of this variety, virtually every national health care delivery system has become more complex over time, particularly through the course of this century.

At the end of the nineteenth century, a two-tier hierarchy was common to most health care delivery systems in the Western world. Primary care was provided by the general practitioner and an amorphous group of semiprofessional and semitrained midwives and nurses (Knox 1982b). In addition, a variety of types of fringe medicine and quackery existed to meet the needs of the bulk of the population who could not afford proper medical care (Knox et al. 1983). Secondary, more specialized care – as far as it existed – was provided from hospitals, mostly general but sometimes specialized (e.g. sanatoria and maternity hospitals). Specialized personnel were, as now, associated with this secondary, hospital-based tier of the health care delivery system. Modifications to this two-tier hierarchy occurred through the twentieth century in response to changes in medical science and technology together with the shifting priorities of professional organizations and governments. These changes and their implications for access to primary health care will be discussed in Chapter 4. The net result of these developments has been an increase in specialization, both of personnel and facilities, arguably an overspecialization in many cases.

In terms of personnel, the increasing specialization of physicians during this century is similar to that which occurred in many other occupations, particularly those which were concerned with professionalization (Fein 1972), although it has been suggested that in the case of doctors this has gone further than necessary, constituting perhaps a form of hyper-development (Fry 1979). The role of professional associations in the development of specialties and super-specialties will be considered in Chapter 4. The momentum for this increasing specialization and its impact on the delivery of health care has been intimately bound up with the increasing emphasis upon hospital treatment based on technical or surgical intervention. Many specialisms grew directly from scientific discoveries (such as the X-ray giving rise to radiology) but, often, specialization developed in medicine with the aim of achieving opportunities in the market and increasing remuneration (Knox et al. 1983). Overall, it is difficult to identify cause and effect in these relationships. However, the result has been to transfer the emphasis from treatment in the home by a general practitioner to treatment in a hospital by highly specialized personnel drawing upon the resources of advanced medical technology. This differentiation of *settings* for medical care is as important as the growth of specialization into a hierarchy, and the changing importance and *relative location* of these settings must be a focus of any consideration of changing patterns of equity or efficiency in health care (Starr 1977; Knox et al. 1983).

This is, of course, a highly generalized outline and a renaissance has taken place in primary care in Britain and other Western countries in the last decade or so, chiefly in response to changing patterns of disease and illness (Hunt 1957; Phillips 1981a). These include the increasing frequency of chronic conditions associated with ageing populations, which are suited to treatment in a home or community environment, versus acute conditions conventionally demanding hospital-based care. Indeed, it appears that through the development of effective treatment for previously acute and deadly conditions, advanced medical technology has contributed substantially toward the increasing importance of care for chronic conditions (Fry 1979). It is notable, for instance, that two-thirds of all mortality in England and Wales in the late 1970s resulted from a trio of chronic conditions: heart disease, cancer, and stroke (Phillips 1981a). This seems to be the pattern of mortality associated with economic development (Kohn and White 1976). However, primary care can also be the only economically viable care available for many developing countries.

Keeping these emerging trends in health care delivery in mind, it is possible to characterize the contemporary health care delivery systems in most developed countries as tripartite. At the primary level are general practitioners, public health clinics and personnel, midwives etc., the general hospital and the specialized physician constituting the secondary level, with specialized hospitals and clinics at the tertiary level. There is some overlap in function among these levels, particularly between the primary and secondary. Ingram et al. (1978), amongst others, have presented evidence for the use of hospital emergency and outpatient units as a substitute for general practitioner services (see Chapter 6). The three-tier hierarchy is favoured over the more traditional two-tier one because it allows a more precise characterization of the aforementioned geographical implications of the changes in health care delivery systems over time.

Developments in medical practice and technology throughout the twentieth century have steadily increased the facility orientation of health care delivery. This is true at the primary level of health care delivery as well as at higher levels, where hospitals have steadily become less 'general' (Haynes and Bentham 1979). As in most service delivery systems, the hierarchical organization of services is mirrored in the spatial pattern and frequency of facilities. Therefore, facilities and personnel offering primary care are more common than those providing secondary care, which, in turn, are more common than those providing tertiary care. A simple diagrammatic representation of this hierarchy is presented in Figure 3.1, a generalization of the country-specific referral systems shown in Figure 2.4. Chapter 2 indicated that these levels are to be seen in developed and developing countries, even if the hospital orientation may be decreasing in the latter. Figure 3.2 illustrates a hierarchy in developing countries, although there is obviously considerable diversity at primary and secondary levels (Fendall 1981). In spatial terms, this health service hierarchy is superficially analogous in form to the central place system familiar to most geographers (Shannon and Dever 1974). However, in terms of operation, the health care delivery hierarchy functions differently from a central place system.

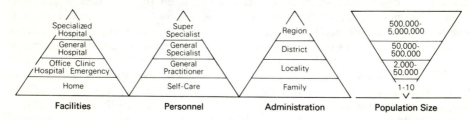

Figure 3.1 The health care hierarchy in terms of facilities, personnel, administration and population served
Source: after Fry (1971)

Functional differences are of two main forms. First, the location of facilities and personnel of different types and levels may be interdependent. Most notably, there is strong evidence that the location of general practitioners is influenced by the location of hospitals providing necessary support services such as X-rays, blood tests, and so on (Phillips 1981a). White (1979) goes as far as to suggest that this locational linkage

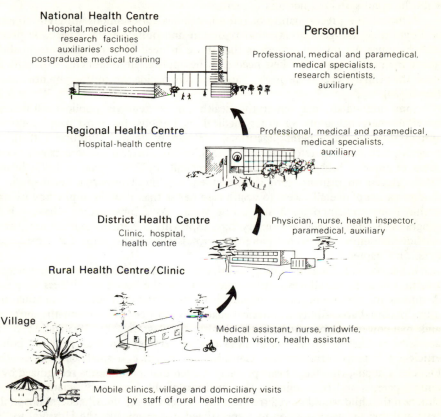

Figure 3.2   A possible health care hierarchy in a developing country.
Based on an idea by Fendall (1981)

among related public facilities should be considered a pre-eminent feature of their spatial organization. At a primary level, however, the distribution of free-enterprise physicians can mirror the pattern of retail facilities (Earickson 1970; Thomas 1976). These propositions, amongst others, are examined in detail in Chapter 4. Secondly, notwithstanding the existence of these important linkages between the locational pattern of facilities and personnel at different levels of the health care delivery hierarchy, the other functional difference between the health care delivery hierarchy and other central place systems is behavioural rather than locational.

Proximity to a general hospital or to a more specialized health care facility does not necessarily imply accessibility. By contrast, proximity to a high-order central place offering highly specialized retail services does. Consequently, the central place analogy breaks down because individuals do not as a matter of course have direct access to different levels of the health care hierarchy, but must instead usually follow a prescribed path through the health care system. These paths are collectively known as the referral system (Shortell 1972). Being referred from one level to another depends, naturally, on the patient's ability to enter the system initially so, again, the analogy with the retail system is not totally realistic.

Although the precise nature of referral systems varies among different countries and over time, the general situation is similar and quite simple. The basic premise underpinning the referral system is that those in need of health care are unable to gauge the precise nature of their need and the appropriate treatment. It usually falls upon the primary level of the health care system to decide this for the potential patient (Fry 1979).

Some individuals may enter the health care system via public health clinics, hospital emergency units, or other medical institutions, but most people enter the system through the general practitioner or 'family doctor' (Phillips 1981a). The general practitioner therefore not only provides basic care, but also acts as a referral agent or filter to higher levels of care (Mechanic 1972; Dartington 1979; Stimson 1980). Access to primary care is consequently a more important element in the determination of overall access to health care per se than is often implied by its status within many contemporary health care systems. As a consequence, changes in the geographical organization of primary care, especially of general practitioners, have a considerable impact on overall patterns of accessibility. This impact is considered in detail in Chapter 4.

In summary, the organization of health care delivery, particularly the prevalence of referral practices, introduces a special element into the health care accessiblity issue. Assuming the absence of socio-economic barriers for the moment, an individual has a given potential accessiblity in physical space to each tier (and the constituent facility and manpower components) of the health care system, but realization of this accessibility (that is, utilization of an 'available' service) may well depend upon the efficiency of the referral system. Consider, for instance, a small child in southern Ontario, Canada, who has a heart problem. Given that a problem is recognized by the child's parents and is brought to the attention of the family physician, in an ideal situation the child would be referred to a cardiologist at the 'local' general hospital for tests and then, if necessary, on to a specialized institution like the Hospital for Sick Children in Toronto for the appropriate, highly specialized treatment. In this situation, the *critical* accessibility is to the general practitioner who initiates the chain of referral. It is largely for this reason, and because of the current WHO emphasis on primary care, that access to general practitioners is taken as the focus of the following two chapters.

Potential versus revealed accessibility

The emphasis of the discussion so far has been upon the factors that may intervene between the recognition of need for health care and the satisfaction of that need. These intervening factors were seen to incorporate at least three interrelated dimensions: the geographical or physical, the socio-economic, and the organizational. However, no reference has yet been made to how the impact of these factors might be monitored, that is, to the nature of relevant indicators. In addition, although most commentators agree on the value of indicators to monitor the delivery of an important service like health care (Aday and Andersen 1974), there is no such consensus on the design of indicators. Broadly speaking, there are two schools of thought. The first advocates a focus on utilization patterns whereas the second concentrates on potential barriers to utilization.

Although the general argument by Donabedian (1973), introduced in Chapter 1, for revealed accessibility (utilization) measures is valid, it masks two important weaknesses. First, 'need' for health care is difficult to define (Boulding 1966). Bradshaw (1972) has gone as far as to suggest that at least four definitions of need are in common circulation in the health care delivery field: 'normative need', 'felt need', 'expressed need', and 'comparative need'. The first three pertain to the need of individuals whilst the fourth refers to that of groups. Normative need is need that is professionally determined for the individual (an apt aphorism might be 'the doctor knows best'). Felt need is need that is perceived by the individual, that is, it is something the individual wants. If a felt need is turned into demand for a service, it becomes an expressed need. Comparative need, like normative need, is professionally determined, and is usually based upon some aggregate expectation of expressed need (that is, demand) in the light of the characteristics of the group. For instance, if it is normal practice for retired people to receive regular general physical examinations, a group of retirees who do not receive this service is in comparative need.

In terms of gauging the extent to which needs are met by the pattern of revealed accessibility, felt need is clearly a difficult concept to operationalize without extensive (and expensive!) household surveys and a sterling faith in the ability of the layman to recognize illness. As a result, many surveys such as the British General Household Survey have to rely on rather unsatisfactory measures of need or ill-health/morbidity indicators, such as number of days' disability from work. These are very difficult to apply realistically or to translate into need because certain occupations, for example, may enable someone only slightly unwell to continue to work, whilst a strenuous physical job may prevent a person working even with a relatively minor illness. Expressed need is impossible to use to gauge the degree of correspondence between need and revealed accessibility because the most common way of expressing need is through utilization of a service. A perfect correspondence between need and utilization would be assured! Therefore, the 'need' referred to by Donabedian must be equated operationally with professional judgement, either at the individual level (normative need) or the group level (comparative need). The objectivity of such a revealed accessibility measure would therefore rest squarely upon that of the professional judgement of need.

Second, there is a multiplicity of factors intervening between the translation of need for health care into use of available facilities. Even if the pattern of need (however defined) is known, examination of utilization patterns does not usually permit the identification of the relative importance of the various geographical, socio-economic, and organizational factors interposed between the need and utilization (Fiedler 1981). This problem will be discussed in detail in Chapter 6 but, nevertheless, it is critical here to appreciate the magnitude of the task. Figure 3.3 presents one interpretation (among many) of the health service utilization process. In this representation, which is derived from Gross (1972), (spatial) accessibility is portrayed as only one element in the equation that translates a negative perception of health level ('need') into utilization of health services. Enabling factors include the socio-economic barriers referred to in the second section of this chapter (income, ethnic status, etc.) and predisposing factors embrace behavioural as well as attitudinal regularities. In Chapter 6, we shall review research that demonstrates that these factors elude easy definition, let alone precise

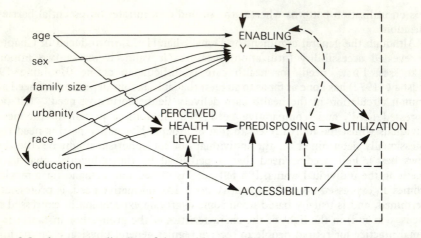

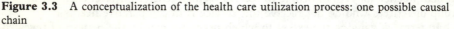

Figure 3.3 A conceptualization of the health care utilization process: one possible causal chain

Source: Gross (1972)

calibration of their impacts on health care utilization.

By comparison with the revealed accessibility approach, the potential accessibility approach is more narrowly focused, the emphasis being upon the opportunity or potential for a certain type of behaviour rather than upon actual behaviour (Moseley 1979). In this, potential accessibility might be determined in terms of the socio-economic and organizational dimensions of society and the health care system respectively, although the emphasis here is on potential physical or spatial accessibility. It is the measuring of this potential physical accessibility that constitutes the focus of Chapter 5. In the following chapter, however, the consideration is of the major regularities evident in the location of primary health care facilities and the factors underlying these regularities.

4　Physician Organization, Location and Access to Health Care

In Chapter 3, we sought to provide a simple conceptual framework for geographical studies of access to health care. The intention here is to consider the access question in one of its two forms, potential accessibility (in contrast to revealed accessibility, which will be considered in Chapters 6 and 7). Following a broad discussion of recent trends in the organization and location of health care, two emphases will be persued. The initial focus will be upon factors influencing the location of physicians, who were earlier noted to be the key resource in health care delivery. We then move to survey evidence drawn from various countries on areal variation in potential access to physician services. In this latter part of the chapter, evidence will be considered without particular reference to the methods and procedures by which it was collected, analysed, and presented. Important questions of accessibility measurement, particularly in the medical manpower context, will be taken up in Chapter 5.

Recent trends in the organization and location of health care

In Chapter 3, it was proposed that the health care systems of developed countries have become increasingly complex in organizational and locational terms through the course of this century. Notwithstanding important differences in national health care policies and resultant systems for health care delivery (Chapter 2), the twentieth century has witnessed the gradual transformation of simple two-tier systems into more complex, multi-tier structures. A three-tier organization is now probably the most typical.

Changes in health care organization have typically been complex, and opportunities for useful generalization concerning them are further limited by the difficulty of attributing locational and accessibility changes to particular organizational developments. In Chapter 3, we were content to identify rapid and cumulative developments in medical science and technology and the differentiation of labour during the process of industrial urbanization as major driving forces behind the progressive specialization of medical personnel and facilities. The locational responses were characterized in broad, central place terms. It is now appropriate further to document organizational changes and to speculate upon their locational and accessibility implications.

Changes in medical science and technology, however ubiquitous they may be, do not require identical responses in terms of the organization and location of health care resources. Ample evidence for this is provided by even the most cursory comparison of the health care delivery systems of the United States and of Great Britain, for instance. The forces shaping organizational and locational responses to improvements in medical science and technology can be divided into three broad categories: economic factors, professional organization, and government intervention.

The importance of economic factors stems from the cost characteristics of medical innovation. It is clear that the adoption of increasingly expensive modes of treatment was, and is, a powerful incentive for centralization of services into increasingly efficient units (in terms of economies of scale to the provider). The most obvious manifestation of the scramble for economies of scale in facility planning has been the running down or closure of small 'community' hospitals, particularly in rural areas (Haynes and Bentham 1979), in favour of larger, more specialized and centrally located (urban) institutions.

Although economies of scale in provision were undoubtedly an important force behind the specialization of facilities through the course of this century, economic imperatives for the division of labour provide only a partial key to understanding the progressive specialization of medical personnel. Knox (1982b) has proposed 'professionalization' of health care personnel, particularly of physicians, to be a key element in the movement toward specialization and, therefore, in the historical evolution of health care systems. In Great Britain, for instance, professionalization of physicians can be traced back to the second half of the nineteenth century and occurred in response to concern over the threat of unorthodox therapeutics and competing paramedical professions *and* over the overcrowding of the orthodox profession, together with a desire for the upward social mobility that would result from an enhanced professional status (Knox 1982b). Knox and Bohland (1983) and Knox et al. (1983) suggest similar considerations to have underlain the drive for professionalization in the United States.

The professionalization of physicians has had several impacts. On one hand, the control of standards within the medical profession through licensing procedures and stricter controls on medical schools was undoubtedly beneficial for the general public. However, concomitant restrictions on the output of medical schools served to reduce the overall accessibility of care. Knox and Bohland (1983) note, for example, that 5,200 physicians graduated from American medical schools in 1900 but only 2,760 in 1918. Additionally, throughout this century, medical associations in various countries have strongly and consistently advocated the benefits to the public of specialization (of physicians) whilst at the same time vociferously condemning the practitioners of less orthodox specialisms such as chiropractic, thereby reinforcing their claim to control over manpower supply and composition.

The third category of forces shaping developments in health care organization and location – government intervention – is even more complex than professionalization. Government involvement in health care delivery is most obviously manifested in the state-run and financed health care systems of countries like the USSR, Sweden, and Great Britain. However, in addition to being a keystone of policy in social democracies, goverment intervention in health care delivery is apparent even in free-market systems such as those of the United States and Australia, and may take the form of medical insurance schemes for the underprivileged, physician location incentives, or medical research programmes. The incentive or justification for government intervention in health care delivery stems from the perceived importance of health care as a service and, particularly in social democracies, from the recognition that the market is an imperfect medium for matching demand and supply in health care. In many ways the increasing effectiveness of medical care throughout this century can be seen as the major incentive for greater government intervention, whilst the increasing complexity of delivery systems progressively opened up greater possibilities (and necessity) for intervention.

It should be obvious even from a discussion as brief as this that the economic, professional, and government intervention factors which have influenced historical, organizational and locational responses to advances in medical science and technology, and which shape contemporary delivery systems, are complex and overlapping. This complexity can be partly controlled for by focusing on a major actor in health care delivery, the physician. This will permit a more thorough discussion of the locational aspects of health care organization than would otherwise be possible.

The discussion of physician location is divided into two parts. The section that follows considers evidence on the factors that influence physician location, both at the level of the individual and in the aggregate. The section following that, on the other hand, considers regional disparities in physician supply that result from the summation of individual location decisions and discusses the implications of such supply disparities for access to health care. Broadly speaking, this sectional breakdown represents a fairly conventional separation of process and pattern. As always, this dichotomy is primarily for convenience in exposition and is not meant to suggest any conceptual independence.

Factors influencing physician location

The task of generalizing about the factors that influence the location of physicians is, on the surface at least, daunting. Physicians are not homogeneous units. As Rogers et al. note,

> Unlike the proverbial rose, a physician is not just a physician. Increasing levels of
> medical knowledge, technology, and training have created a medical care system in which
> specialization is the norm rather than the exception.
> (Rogers et al. 1980, p. 45)

It is probable that the factors affecting the location of physicians will differ systematically by specialization. Moreover, the nature of these variations in locational determinants varies considerably over time and amongst health care systems. However, it remains necessary to seek generalization about physician location, partly because of the overall importance of the physician in health care delivery systems, but also because the individual physician is generally the most volatile component of the locational infrastructure of health care delivery. By contrast with hospitals and clinics, with their substantial capital investment, individual physicians are highly mobile. This mobility creates a potential for rapid and extensive changes in supply over space, even if locational inertia has been a notable characteristic of physician distribution in many systems.

A review of the literature on the determinants of physician location reveals the usual variety in orientation and emphasis that stems from the disciplinary origins and personal predilections of the commentators. Setting those aside, there appear to be two mutually exclusive approaches to investigation, one *ecological*, the other *behavioural*. The ecological approach poses questions such as: 'What sorts of characteristics are associated with areas with a favourable ratio of physicians relative to population?' The behavioural approach, by contrast, seeks to establish the nature and relative importance of factors relevant to the locational decision-making of individual physicians.

The ecological approach

Two types of work will be considered here. The first has as its explicit focus the 'explanation' of aggregate patterns of physician distribution, with the preferred form of investigation being multiple correlation and regression analysis. The second element within the ecological approach is characterized by a primary focus on the measurement of disparities in the geographical distribution of physicians, the 'explanation' of observed disparities following as a natural complement.

The correlation- and regression-based investigation of factors influencing the location of physicians has been pursued most vigorously in the United States, with the spatial unit of analysis ranging from the state (Benham et al. 1968) to the urban census tract (Elesh and Schollaert 1972). In each case, the intent has been to calibrate the relationship between the distribution of physicians by area and the characteristics of those areas. In multiple correlation and regression terms, this translates into a linear model of the following structure

$$P = a \pm b_1 X_1 \pm b_2 X_2 \pm \dots b_m X_m$$

where the X variables are the independent or causal variables proposed to influence variation in P, the dependent variable (in this case, the number of physicians). The partial regression coefficients(bs) measure the elasticity of the dependent variable to changes in each independent variable, all others being held constant (Yeates 1974; Taylor 1977a). When standardized in terms of the standard deviations of the dependent and independent variables (which removes the effect of different measurement scales amongst the independent variables), these coefficients can be interpreted as measures of the relative importance of causal variables.

American studies of the ecology of physician distribution have consistently reported population size to be the most important single correlate of physician supply. Whether by state (Benham et al. 1968) or by metropolitan/urban area (Marden 1966), the number of physicians in an area has been found to be positively correlated with its population. However, this relationship varies in strength depending upon the definition of the units of analysis, both of the dependent variable (physicians) and the independent variable (population). By analysing 1960 data for 369 Standard Metropolitan Statistical Areas (SMSAs) and large urban areas in the United States, Marden obtained extremely high correlations between physician numbers (total, general practice, and specialized) and population size (Table 4.1). However, on disaggregation by city size category, greater variations in the covariance of population and physician numbers were revealed. The strength of the population effect clearly declines systematically with decreasing city size and, in most cases, the correlations for total number of physicians are greater than for general practitioners which, in turn, are higher than those for specialists. This suggests that these ecological correlations should be interpreted with considerable care. Clearly, given that one would expect different relationships between population size and number of specialists, and population size and number of general practitioners (a non-linear positive relationship for general practitioners), the generally higher correlations for *total* physicians than for either of the component groups suggest that the former may be spurious, a misleading 'average' effect. Marden himself suggested that the decrease in all three correlations with decreasing size class might well be a consequence of the decreasing range of city size within each category (500,000 for the second but only 50,000 for the fifth). It seems that city population size is even more important in developing countries, particularly where primary prevails and, in some countries, over 75 percent of doctors may be in larger centres although they may contain as little as one-quarter of the total population.

Table 4.1 *Correlation between population size and physicians in private practice, by type of practice for six size categories, metropolitan United States, 1960*

Population size category	No. of cases	Total group of physicians in private practice	Private practitioners by type of practice	
			General	Specialized
All cases	369	.98***	.99***	.97***
1,000,000 + ...	24	.98***	.99***	.96***
500,000–999,999	33	.85***	.80***	.74***
250,000–499,999	47	.81***	.62***	.64***
100,000–249,999	123	.81***	.73***	.67***
50,000– 99,999	111	.60***	.58***	.41***
Under 50,000	31	.51**	.51**	.29

Significant at the 0.01 level   * Significant at the 0.001 level

Source: after Marden (1966)

Acknowledging that population size is by far the most important ecological correlate of physician supply, it remains to assess the importance of other variables, related perhaps to the demographic and socio-economic composition of populations. Analytically, this can be achieved in one of three ways. The most obvious route to follow would be to include explanatory variables additional to population size in a multiple correlation and regression model. Although it affords an opportunity to compare directly the importance of the population size variable with that of other independent variables, the practicality of this option is severely undermined by the general lack of statistical independence between population size and variables such as per capita income and population density. This problem of covariance amongst the *independent* variables is called multicollinearity, the increase of which makes it progressively unreasonable to interpret the standardized partial regression coefficients as measures of the relative importance of independent variables. Results produced in this sort of analysis are, typically, ambiguous (for example, see Benham et al. 1968).

The two remaining strategies for assessing the correlation between physician numbers and population composition variables both rest upon the control of the size effect but differ in the completeness of this control. The partial approach to control is typified by the work of Marden (1966), who developed separate multiple regression models for each of his six city size categories. The four independent variables used were '% population under 5 and over 65', 'median school years completed', '% population that is non-white', and 'number of available beds in short-term, non-federal hospitals'. The results of this analysis are summarized in Table 4.2. The standardized partial regression coefficients suggest that, partially correcting for population size factors, the distribution of general practitioners was most highly correlated with the racial composition and age structure of populations. 'In both instances, the relationships are in the direction of the hypotheses, as the distribution of general practitioners is directly associated with the "high medical risk" group and inversely associated with the proportion of non-whites in the population of a metropolitan area'

Table 4.2 *Relative importance of compositional and environmental factors in multiple correlation analysis, physician distribution, metropolitan United States, 1960*

Rank	1,000,000 +	500,000–999,999	250,000–499,999	100,000–249,999	50,000–99,999	Under 50,000
			Total Physicians			
1	Race −.21	Educ .40	Beds .25	Age .32	Age .39	Beds .35
2	Age −.20	Beds .36	Age .20	Beds .32	Educ .25	Race −.34
3	Educ .14	Race −.32	Educ .20	Educ .25	Beds .23	Age .28
4	Beds .12	Age −.18	Race −.16	Race −.13	Race −.09	Educ −.04
			General practitioners			
1	Race −.41	Race −.66	Race −.40	Race −.49	Age .49	Age .54
2	Age −.22	Educ .24	Age .26	Age .36	Race −.15	Beds −.37
3	Educ .12	Age −.07	Educ −.21	Beds .13	Educ .07	Race −.35
4	Beds .06	Beds .07	Beds .06	Educ −.09	Beds −.02	Educ −.10
			Specialists			
1	Age −.18	Beds .38	Educ .42	Educ .37	Beds .32	Beds .70
2	Beds .14	Educ .34	Beds .23	Beds .29	Educ .27	Race −.18
3	Educ .14	Age −.17	Race .19	Race .15	Age .13	Age −.04
4	Race −.11	Race −.01	Age −.02	Age .12	Race −.01	Educ .02

Source: Marden (1966)

(Marden 1966, p. 299). By contrast, the distribution of specialists was most strongly associated with the availability of hospital support facilities and the educational status of the population.

By contrast with the partial approach to controlling for the population size effect, Joroff and Navarro (1971) totally controlled for population size by using physician: population ratios as dependent variables in their analysis of physician distribution across 299 American metropolitan areas in the mid-1960s. They found the best predictor of general practitioner ratios to be the percentage of the population over 65, whereas the distribution of specialists was most strongly correlated with median level of education and general hospital bed rate (per 1,000 people). Except for the failure to identify a significant correlation between the distribution of general practitioners and the racial composition of city populations, these results are similar to those of Marden (1966).

Rather than cataloguing the results of further studies, attention will instead be drawn to some of the problems associated with the ecological approach to the explanation of physician distribution. The fundamental limitations of this approach lie in the area of research design, whilst additional problems stem from the technical demands of correlation and regression as a framework for analysis.

Ecological analysis necessitates categorization, and therefore space must be divided into discrete regions whilst physicians are usually grouped by type. The regionalization problem is all too familiar to geographers, who constantly face the challenge of assessing the impact of modifiable areal units and of scale on empirically generated results (Taylor 1977a). In connection with this, it has already been noted that Marden (1966) and Benham et al. (1968) chose to divide the USA into regions along totally different lines, respectively by metropolitan/urban area and by state. Additionally, studies have differed in their definition of the dependent variable. All studies, explicitly or implicitly, assume that all physicians of a given type deliver an identical quantity and quality of service – an assumption which will be critically evaluated later in this chapter (see 'Micro-studies of physician availability') and in Chapter 8 – but while Marden, for example, treats specialists as a homogeneous subgroup of physicians, Joroff and Navarro (1971) disaggregated by specific specialty. Indeed, the relatively poor explanatory power of Marden's specialist regressions (Table 4.2) may well be related to the heterogeneity in his dependent variable.

The choice of independent variables is almost equally important in influencing the results of ecological analysis. The normal practice in ecological analysis is to postulate the major influences on physician location and then to identify variables to represent them within the correlation and regression framework. There is a possibility for choice therefore in *both* the identification of factors to be included in the analysis *and* the way in which they are to be represented. Setting aside the easily appreciated problem of choosing between alternative data series that measure the same broad factor (for example, hospital beds per 1,000 people versus per capita expenditure on hospitals), the fundamental issue lies in identifying the appropriate set of factors to be represented in the multiple correlation and regression analysis. Substantial differences in the set of factors represented in analysis can make constructive comparison almost impossible. While Rimlinger and Steele (1963), for example, found a significant positive correlation between the number of physicians per capita and per capita income across the United States, Marden (1966) did not include an income variable in his analysis! Another manifestation of this same problem relates to the degree to which factors are effectively included in an analysis or simply incorporated in a token manner. For instance, Benham et al. (1968, p. 339) acknowledged that 'It is widely believed that medics have strong locational preferences, preferring to be near hospitals and other facilities. Like other professionals, they desire to locate where cultural facilities (schools, theatres, etc.) are available. Also, they presumably exhibit a normal degree of inertia that, ceteris paribus, leads them to prefer to remain where they were raised and/or educated....'. Nevertheless they chose to represent these non-pecuniary factors by only one variable, 'volume of training facilities'! An important problem in cross-cultural comparisons is, however, that identical variables are rarely available, and the income factor per capita is perhaps one of the most elusive.

On the technical side, the major problem associated with the use of multiple correlation and regression analysis is multicollinearity, although the reliability of results may also be adversely affected if other assumptions, such as linearity and normality, are not met (Farrar and Glauber 1967; Poole and O'Farrell 1971; Johnston 1978). The identification of muticollinearity as a problem of special relevance to the ecological analysis of physician distributions stems from the explicit intent in ecological studies to evaluate the *relative importance* of independent variables and hence the factors they represent. It is this very ability that is compromised by substantial levels of interdependence or covariance amongst the independent variables; even at relatively 'low' levels, the presence of multicollinearity forces a conservative interpretation of partial regression coefficients.

Interpretation of the results of any investigation of the covariance over space of physician numbers and selected descriptors of populations has therefore to be carefully phrased in the light of the research design and the technical limitations of the analytical framework (Phillips 1981a). Additionally, there remains the ever-present problem of ecological fallacy (Robinson 1950; Gudgin 1975). Even the most methodologically and technically rigorous empirical analysis of ecological data does not protect the researcher from false inferences about individual behaviour. In particular, it is extremely difficult to differentiate from ecological analyses amongst the various behavioural and organizational factors that might be simultaneously influencing the location of medical manpower. An observed tendency for physician supply to become more sensitive to the aggregate income characteristics of populations might be attributed to several discrete behavioural factors, all of which are equally plausible (for example, an increasing preference for practice amongst peer groups versus a decrease in government-sponsored insurance programmes for the underprivileged). It is only at the level of the individual that wholly reliable observations about behaviour can be made. However, this does not invalidate aggregate analysis. Instead, it may be suggested that aggregate analysis should be restricted to statements of pattern and that their interpretation in relation to processes should be validated at the level of the individual.

As mentioned earlier, studies whose primary aim has been to identify major regularities in physician supply have frequently suggested reasons for observed disparities. In most cases, this is in the spirit of the final statement of the preceding paragraph. For example, observed aggregate disparities in the urban–rural supply of doctors have been related to preferences on the part of individual physicians for urban practice and/or to changes in the organizational structure of health care delivery that have emphasized the advantages of the urban treatment locus. Given that these studies of disparities are dependent upon an appreciation of behavioural regularities and constraints upon them, further discussion of them will be postponed until later in this chapter (see 'Areal variation in physician availability').

The behavioural approach

In this section, observations on the locational decision-making of physicians are placed within the context of broad, overlapping changes in health care delivery systems, professional organizations, and government policy. Some studies of locational behaviour have been essentially anecdotal, based perhaps upon the personal experience of the author or on observations of peer groups, whilst others have been more rigorously grounded in the traditions of social science research on individual behaviour and preferences. Although the former will not be ignored, the attention here will be on examples of the latter type of study.

The distribution of physicians within any geographically defined health care system is the aggregate outcome of decisions made with respect to individual preferences within a set of constrained choices. In considering the major features of these preferences and of the various constraints on locational choice, the initial focus will be on the preferences of individuals.

(a) Individual preferences It is axiomatic that individual preferences are a function of attitudes (Fishbein 1967). In the context of the locational choice of physicians, it is possible to identify three major elements within attitude formation: personal, professional, and class or lifestyle. The personal component of attitude is that part which is unique to the individual and therefore defiant of generalization. Indeed, it is this component which prompts behaviour that may, in a broad context, appear irrational or unpredictable. We shall focus here on those elements of attitude, based upon professional considerations and biases and upon class-based or lifestyle-related aspirations, for which it is feasible to make plausible generalizations.

It has been repeatedly suggested that physicians make their key career choices early. In their survey of physicians in western New York State, Parker and Tuxill (1967) found that 42.3 percent of physicians practising in large communities and 37.7 percent of those practising in small communities had made their career specialization and locational choices in their internship/residency periods. A further 20.0 percent and 26.4 percent, respectively, reported that they made these choices in early years of practice. A later study, by Diseker and Chappell (1976), found the locational priorities of residents and practitioners to be virtually identical. Taken together, these studies point toward the importance of the professional views espoused by mentors, teachers, or peers in the medical school environment. These professional views will tend to differ among countries, of course, and may also change over time, but they often feature some value bias toward specialization rather than general practice, and the notion that access to an adequate range of medical facilities is a prerequisite for proper medical practice, at least in part a function of the 'scientific' orientation of Western medical training. Parker and Tuxill note that in locational terms this translates into a bias toward practice in larger, urban communities, the population of smaller, perhaps rural communities being inadequate to support (desirable) specialization or ('needed') back-up facilities.

The bias toward practices in larger communities may be reinforced by class or lifestyle-related considerations. Parker and Tuxill (1967), for instance, noted that 60.4

percent of their sample subgroup who were practising in small communities originated from smaller communities and that 73 percent of their large-community physicians came from large communities. They suggest that this bias toward a familiar physical and social environment can, in its most extreme form, place a high priority on proximity to relatives and friends and may be stronger on the part of the spouse than the physician. The importance of spousal attitudes and preferences is reported also by Diseker and Chappell (1976).

Class and lifestyle attitudes may be directly manifested in a preference for location in an area of suitable (peer) social status and/or in a preference for practising in areas with good access to leisure or recreational activities (Knox and Pacione 1980). Many authors have pointed out that physicians come very predominantly from the higher social classes (in Britain, from Social Classes I and II). Bridgstock (1976), for example, reported that 87 percent of doctors in a survey in Britain came from non-manual occupation homes. Such findings have implications for the type of lifestyle which doctors expect. However, in overall terms, lifestyle attitudes may achieve their greatest locational impact indirectly via preferences for a maximum separation of work and leisure time. Both Diseker and Chappell (1976) in the United States and Knox and Pacione (1980) in the United Kingdom reported the availability of practice coverage to be an important locational priority for physicians. Indeed, this factor has been viewed as critical in accounting for the problems of recruiting physicians into small, mainly rural community practices in the United States (Bible 1970).

> One doctor in an isolated western area, having responded to patients until he was exhausted, took the telephone off the hook and fell into bed at midnight. He and his family were soon awakened by a loud banging on the door. A patient had spotted his car in the garage and wanted him to see a sick child.

Thus Bernstein et al. (1979, p. 5) describe the predicament of the single-handed rural practitioner. It is not difficult to appreciate the attractiveness of practising in situations where the delivery of health care can be organized to meet the needs of the health care provider as well as the patient! Only in larger centres, for example, will commercial out-of-hours 'deputizing' medical firms work, or only in such centres will there be sufficient numbers of doctors to be able to share out-of-hours work.

To round off this discussion of individual preferences, we should like to draw attention to one of the most thorough studies to date of locational decison-making by physicians, reported by Knox and Pacione (1980). They surveyed students in their final two years of medical school in Dundee and Glasgow, Scotland. Respondents were asked to assess the potential importance of a predetermined set of factors in their choice of work location and to indicate preferred place of practice, by region within Scotland and by neighbourhood within the city in which their medical school was located.

The major locational factors cited by the Scottish students are broken down in Table 4.3. Using an arbitrary cut-off of 20 percent for each subgroup as a measure of 'significance', the results support the earlier contention regarding the importance of professional and lifestyle-related regularities in the formation of preferences. Interestingly, financial incentives appear to play only a minor role in the decision-making lexicon of these medical students. Indeed, Knox and Pacione found that, when asked to suggest areas in Scotland that had been designated as 'underdoctored' (therefore eligible for special grants) or 'overdoctored' (with restrictions on entry for new practitioners), 'between a quarter and a half of the respondents could offer no

Table 4.3 *Major reasons given by medical students for choosing work location (Glasgow and Dundee medical schools)*

Reason	(%) mentioning as important		(%) mentioning as not important	
	Glasgow	Dundee	Glasgow	Dundee
Access to leisure/recreation facilities	43.8	50.0*	5.7	5.9
Proximity to relatives/friends in home area	20.0	23.5*	10.5	23.5
Accessibility to specialized hospital facilities	57.1	50.0*	1.9	5.9
Opportunity to treat private patients	2.9	5.8	62.9	61.7
Availability of suitable surgery/office space	32.4	26.4*	3.8	11.7
Opportunity to build up an extensive list of NHS patients	8.6	2.9	11.4	38.3
Opportunity to work in a rural practice	39.1	55.9*	12.4	8.8
Availability of manpower to cover illness, vacation and time taken off for in-service tranining	46.7	58.9*	1.0	5.9
Opportunity to work with a particularly 'needy' population (e.g. in an immigrant or inner-city area)	16.2	11.7	23.8	26.4
Proximity to your spouse's family	0.0	2.9	60.0	26.4
Proximity to your old university medical school	0.0	0.0	76.2	41.1
Eligibility for financial incentives	6.7	2.9	10.5	11.8
Opportunity to work in a Health Centre	25.7	8.8	16.2	32.3

*Affirmation rate in excess of 20% for both sample groups

Source: after Knox and Pacione (1980)

suggestions whatsoever' (Knox and Pacione 1980, p. 48). Moreover, despite the suggestion by Lankford (1971) and De Vise (1973), among others, that the profit motive is of primary importance in the locational decision-making of physicians in the United States, Diseker and Chappell's (1976) analysis of practitioners in the American south-east suggests a partial indifference to economic motives similar to that found amongst the Scottish students surveyed by Knox and Pacione. In Britain, incomes of doctors in the public as opposed to the private sector are little influenced by location since work load and list size do not play such a large part as they did previously in determining GP remuneration and locational incentive payments are, overall, relatively small. Additional American evidence is provided by Bernstein et al. (1979), who drew attention to small towns that failed to attract a physician despite offering, among other things, a guaranteed minimum income and adequate free office space.

Knox and Pacione found that locational priorities suggested in Table 4.3 were mirrored by students' regional preferences (Figure 4.1):

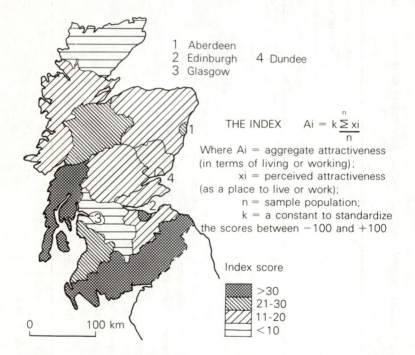

1 Aberdeen
2 Edinburgh   4 Dundee
3 Glasgow

THE INDEX $A_i = k \dfrac{\sum\limits^{n} x_i}{n}$

Where A_i = aggregate attractiveness
(in terms of living or working);
x_i = perceived attractiveness
(as a place to live or work);
n = sample population;
k = a constant to standardize
the scores between -100 and $+100$

Index score

>30
21-30
11-20
<10

0   100 km

Figure 4.1 How medical students view places in which to work: the regional preferences of potential Scottish general practitioners

Source: after Knox and Pacione (1980)

In general, the preference surface exhibits a fairly even plateau, with a depression corresponding to the industrialized belt of the central lowlands (where most of the officially designated underdoctored areas are) and a tendency to shelve off toward the remoter rural areas of the northern highlands.
(Knox and Pacione 1980, pp. 49–50)

At the intraurban level, both residential and work place preferences were highly localized and differentiated, with a constant avoidance of the inner city and of public housing areas.

The preference mapping attempted by Knox and Pacione represents only the degree to which different areas are *perceived* to meet the broad locational requirements of individuals. The actual set of opportunities available to individuals will depend upon the nature of the health care delivery system *and* the profession of which they become a part and the way that both the system and the profession serve, directly and indirectly, to constrain locational choice.

(b) Constraints on locational choice   Medical personnel are part of an overall organization, the health care delivery system, which is more or less tightly regulated. The breadth of opportunity for individual choice is delimited by the parameters of the system in which choice is being made. In the first section of this chapter, it was noted that although advances in medical science and technology provided the general impetus for the progressive reorganization of health care delivery systems around specialized and capital-intensive facilities, the precise form of these new systems was strongly influenced by the intervention of professional organizations and of government. By extension, the policies and actions of these actors have influenced and constrained the locational choices of physicians.

As noted earlier, the professionalization of physicians ranks as a major feature of the development of health care delivery through the course of this century (Knox and Bohland 1983). The importance of professionally based attitudes and preferences in career/location choice has already been remarked upon but the influences of the medical profession do not stop there. In the majority of Western countries, the medical profession has been permitted considerable powers of self-regulation, which in most cases embraces the control of admission to the profession, that is, avoidance of oversupply! In addition, through their direct or indirect influence on medical schools, professional organizations can also shape the mix of physicians by specialty.

In terms of access, the control of supply is clearly of fundamental importance in that it is fairly safe to conclude that, whatever their precise location, more doctors provide greater aggregate access to care, and fewer doctors lower aggregate access to care. The access implications of controlling entry into specialties arise from the changes in the threshold requirements of specialties that would occur if their supply was altered. Consider, for instance, a hypothetical country of 10 million people. If the medical schools of this hypothetical state produced dermatologists at an annual rate that ensured a constant supply of 100 practitioners, dermatology would be a high-order specialization, each dermatologist catering for the needs of 100,000 people on average. In central place terms this translates into a locational bias toward high-order places, namely large cities. However, if this hypothetical state decided to increase the production of dermatologists to a level that would ensure a steady state supply of 1,000, dermatology would be reduced to the status of a lower-order specialty and its overall locational pattern would adjust accordingly. Thus, the medical establishment not only influences the attitudes of its members toward individual specialties but, through the control of entry into them, also restricts potential for dispersion within the organizational-locational hierarchy of the health care system.

It would be unwise, however, to assume that professional control over specialization is premeditated or even conscious although it may well be so for control of overall supply. A Canadian example illustrates this quite well. During the late 1960s, there was a general perception amongst Canadian medical professionals that a crisis existed in terms of health care delivery and that this crisis resulted from the failure of manpower supply to increase at the same rate as demand for medical services (Bean 1967). At the same time, however, there was little consensus on the structure (by specialty) of the undersupply. For example, Goldberg (1967) suggested that an increase in the supply of specialists should be given priority, whereas Perkin (1967) favoured general practitioners. This lack of consensus in part reflected general problems of evaluating medical manpower distributions (Hadley 1979; Gray 1980) but, perhaps more fundamentally, the difficulty of anticipating the implications for manpower of (unpredictable) changes in medical science and technology (Kerr 1967). The Canadian response to their 'manpower crisis' was to increase the supply of physicians of all types, to such an extent that a decade after the original crisis was perceived, there appeared to be a general situation of oversupply (Gray 1980)! It is suggested that a similar picture is to be seen in other countries such as Britain and the USA although, of course, oversupply exists only in terms of current workload norms for doctors.

The role of government in constraining the locational choice of physicians is even more pervasive and far-reaching than that of professional associations. This role manifests itself directly through administrative controls on the location of practices and/or locational incentives for physicians and indirectly through government control (or partial control) of various segments of the health care delivery system. Consider the indirect role first.

In most Western societies, government involvement in health care delivery is complex and far-reaching. In the majority of instances it is government that determines (through action or inaction) the autonomy of the medical profession. Despite instances where governments have sought to exercise more direct control over the supply of medical manpower, as occurred in Canada in the late 1960s and early 1970s when the Federal and provincial governments moved to expand medical schools, the overwhelming tradition has been to permit a substantial degree of autonomy on the part of professional associations. In national terms, far greater variability has been apparent in connection with policies concerning payment for treatment.

The British National Health Service, perhaps the model for 'free' health care systems, dates only from 1948. Other countries have taken a while to follow the British lead (Chapter 2). In Canada, universal health insurance coverage was introduced province by province during the 1960s (Bryant 1981) and Australia has flirted with nationalized health care through the 1970s (Stimson 1981). Similarly, the United States experimented with 'Medicare' and 'Medicaid' during the 1960s and 1970s (Pyle 1979). The implications of these schemes for the location of physicians lie in their elimination or partial elimination of economic barriers to use of medical services. Given that the geographical distribution of rich and poor is highly regular in all societies, the institution of subsidized or free health care has the potential to enhance the access of 'poor' areas to physician services. However, this potential may not always be fulfilled, partly because physicians may use public insurance schemes to support financially a situation of oversupply. Additionally, the regulations embodied in financial support schemes may themselves perpetuate or exacerbate disparities in health care availability. In the American case, Reilly et al. (1980) partly attribute the problem of attracting medical manpower to rural areas to the practice in Medicare and

Medicaid of reimbursing rural physicians at a lower rate than their urban counterparts for identical services.

A third form of indirect government influence on physician location derives its importance from the increasing facility orientation of health care. Given the known 'attraction' of hospitals for physicians (Knox and Pacione 1980), government policy towards the structure of hospital systems, and especially towards the location of hospitals, has obvious implications for constraining physicians' choice of location. In many cases, the effect will be local: a physician chooses to locate in town A and not in neighbouring town B because A has a general hospital and B does not. However, there may also be more general, systematic effects resulting from policies that encourage increasing hospital size to take advantage of economies of scale. In the United Kingdom, for example, the erosion of the role of the cottage hospital within the facility network threatened serious implications for the practice of medicine in small communities (While 1978), particularly those in rural regions (Haynes and Bentham 1979).

Even nearer to the heart of the locational issue has been the growing tendency of government to encourage general practitioners to become facility-based. Using a British example again, health centre policies initiated mainly in the 1960s were motivated largely by a desire to improve primary care services by providing an environment for the economically efficient use of ancillary manpower and specialized equipment and the development of primary health care teams (Royal College of General Practitioners 1979). However, their most important result has been to accelerate already existing tendencies toward centralization of GP services (Sumner 1971). This centralization has been extensive and swift, as is suggested by the West Glamorgan example depicted in Figure 4.2.

Overt government policies aimed at influencing the location of physicians usually fall into one of two categories: administrative controls or financial incentives. Administrative controls are generally of a negative sort – delimiting where physicians should *not* locate. The British NHS system of 'designated', 'open', 'intermediate', and 'restricted' general practice areas is an excellent example of largely negative control with limited positive inducements. The categories are defined in terms of average list size (in Britain, residents register with a general practitioner; that is they are put on his or her 'list'). General practitioners are effectively barred from locating in restricted areas, even as replacements for outgoing physicians. Setting up practice becomes progressively easier as one goes from intermediate to open (where there are no substantial barriers), and for the designated area administrative ease is complemented by two financial incentives: an 'initial practice allowance' and a 'designated area allowance' (Butler and Knight 1976; Phillips 1981a).

The British combination of negative controls and financial incentives is, however, relatively rare. In most countries, attempts to influence directly physician location have been most often based upon financial incentives. An excellent example is provided by the 'Programme for Underserviced Areas (Physicians)' instituted in Ontario, Canada, in 1969 (Bass and Copeman 1975). The financial incentives incorporated into the programme were of two types: a location incentive grant (similar to the British 'initial practice allowance') and a guaranteed minimum annual net income. In addition, bursaries were made available to medical students who agreed to practise in underserviced areas (one year of service for each bursary year).

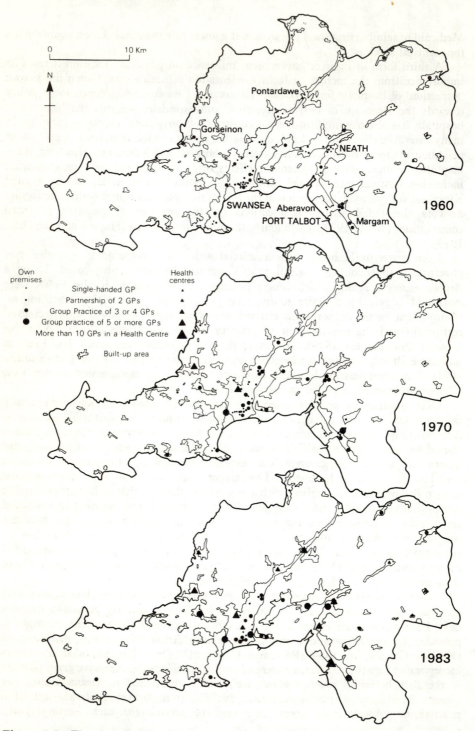

Figure 4.2   The changing organization and location of general practitioner services in West Glamorgan, 1960–1983

Coincidentally, in 1969, a rural incentives scheme for doctors was introduced in New Zealand (Barnett and Sheerin 1977). Under the New Zealand scheme, physicians locating in 'designated' rural areas were given a 10 percent bonus above and beyond regular fee-for-service rates, a rural practice grant (contingent upon two years' service in a designated rural area), assistance in the employment of registered nurses and locums, and more!

By contrast with the British, Canadian, and New Zealand examples cited above, schemes for influencing the location of physicians in the United States have been more diverse, primarily because of the weaker role of central government (De Vise 1973). Indeed many of the 'actors' in the United States are not government agencies at all. The Kellogg Foundation, for example, has underwritten numerous internship and residency schemes in underserviced areas (Reilly et al. 1980). More typical, however, are the efforts of small communities without physicians to attract a doctor – 'Physician Wanted' or 'Doctor, Please Stay' – through income guarantees, student bursaries, free accommodation, and so on (Bernstein et al. 1979). Readers are referred to Eisenberg and Cantwell (1976) for a fuller discussion of American schemes.

Retrospective

The locational preferences of physicians display considerable regularity. This predictability stems primarily from two sources: the professionalization of physicians and their increasing locational orientation to health care facilities. Both are at least in part responses to changes in medical science and technology. It would appear that these two processes have occurred in parallel and have unfolded in such a way as to reinforce mutually their impacts on locational patterns.

The professionalization of physicians has institutionalized entry into the medical profession and formalized the relative standings of specialties. Moreover, it has elevated 'professional' attitudes to a pinnacle of importance. What individual physicians may perceive to be their *individual* attitudes and preferences are to a large degree *peer* attitudes and preferences. More subtly, professionalization can be seen at the roots of a desire amongst physicians to maximize their personal and career goals as opposed to those of society, which emerges strongly in the responses of the Scottish medical students polled by Knox and Pacione (1980). This is an important paradox. Individuals (doctors) are permitted to make (often irrational) locational decisions that impair the effective delivery of a key service. Professionalization has provided the political power base for the perpetuation of this value system (De Vise 1973).

The importance of facility orientation as a constraint on physician location, especially of specialists, has already been stressed. However, for general practitioners at least, one could view this increasing facility orientation as partly voluntary, that is, as going beyond levels necessitated by new modes of treatment. The emergence of group practice as the dominant form of manpower organization for the delivery of primary care could perhaps be viewed as symptomatic of a desire for greater control of 'free time' on the part of general practitioners as much as a means of improving service delivery. Indeed, there is little evidence that the productivity of doctors or the nature of the service delivered is significantly different following the formation of group practices (Evans et al. 1973). Data for England indicate a marked trend not only towards group practice, but towards larger group practices (Table 4.4). Combined with changes in consultation practices (namely, a decline in 'house calls' or domiciliary

L.I.H.E.
THE HARKLAND LIBRARY
STAND PARK RD., LIVERPOOL, L16 9JD

Table 7.4 *General medical practitioners in England: analysis by organization of practices*

Type of practice	1961(%)[b]	1971(%)[b]	1976(%)[b]	1981(%)[b]
all practitioners (total)[a]	20865	20597	22015	24359
unrestricted practitioners (total)[a]	18905	19374	20551	22304
solo doctor practices	5337 (28.3)	3954 (20.4)	3494 (17.0)	2990 (13.4)
partnerships of 2 doctors	6384 (33.8)	4552 (23.5)	4184 (20.4)	4004 (18.0)
groups of 3 doctors	4008 (21.2)	4911 (25.3)	4971 (24.2)	5132 (23.0)
4 doctors	1984 (10.5)	3232 (16.7)	3784 (18.4)	4255 (19.1)
5 doctors	715 (3.8)	1490 (7.7)	2250 (10.9)	2940 (13.1)
6 or more	450 (2.4)	1235 (6.4)	1863 (9.1)	2983 (13.4)

Note: a totals differ as restricted principals, trainees and assistants have been omitted; b percentages are calculated in relation to the total number of unrestricted practitioners.

Source: Phillips (1981a); Department of Health and Social Security (1982)

visits), the rise of the group practice has radically changed the image of the general practitioner service. Bridgstock (1976) in Britain and Parker and Tuxill (1967) in the United States found group practice to be the desired norm for younger, better qualified, and more ambitious physicians.

The increasing facility orientation of physicians has amplified an already existing urban structure effect (Sumner 1971). Put simply, physicians need office space, and as the optimal size of facilities increases at all levels of specialization, the set of feasible sites decreases. In the United States and Australia, the distribution of suburban commercial nodes has been noted to be an important influence on physician location (Miller 1977; Stimson 1980), while Knox (1978) has posited a strong link between the underdoctoring of suburban housing estates in Britain and the limited availability of suitable surgery accommodation. In the USA, there is the locational paradox of need for larger offices (surgeries) yet a desire to remain near to CBD shopping areas which patients attend and where offices (and, hence, fees) are more expensive.

With these observations in mind, what then of overt government attempts to influence physician location? To a considerable extent these have appeared to have been failures because disparities in manpower availability have not disappeared (Eisenberg and Cantwell 1976; Barnett and Sheerin 1978; Knox 1979a; Knox and Pacione 1980). This failure has resulted from the already noted reluctance in most countries to institute firm controls on physician location and from the emphasis within incentive programmes on the *financial* returns from practice. It may be suggested that evidence now points away from financial returns as a primary determinant of physician location. Consider, for instance, that of the 217 students who accepted bursaries during the first five years of the Ontario Programme for Underserviced Areas (Physicians), only 53 percent honoured their contracts, the remaining 47 percent committing themselves to repay their grants. Therefore, nearly one-half of a group that made an initial commitment to take advantage of the financial incentives offered for practice in underserviced areas ended by trading off those incentives against other considerations (Bass and Copeman 1975). De Vise (1973) and Barnett and Sheerin

(1978) go as far as to suggest that physicians, through the power to generate demands for their services, may be able to minimize the financial implications of forgoing such locational incentives.

It therefore seems that policies intended to influence the location of physicians must be modified to take into account known regularities in physician attitudes and preferences and the structure of the physical and organizational systems in which locational choice is effected. Indeed, there are signs that such a policy change may already have begun. In New Zealand (Barnett and Sheerin 1978) and in the United States (Reilly et al. 1980), programmes to introduce medical students to the realities (as opposed to the stereotypes) of rural practice have produced encouraging results.

As well as attempting to wean individual doctors away from learned (professional) attitudes, recent attempts to influence physician location have advocated conscious manipulation of the facility system. In the United States, for example, attempts have been made to make practice in small isolated communities more attractive by more fully integrating urban and rural delivery systems within a regional framework (Moscovice et al. 1979) The success of these attempts is, however, founded upon determination on the part of those who decide on the location of facilities. Barnett and Sheerin (1977) note that, in its first four years of operation, a health centre programme initiated in New Zealand in 1973 had not served to redirect manpower toward underserviced areas. This leads conveniently to a final observation on the location of medical personnel. It remains by and large highly structured but not by conscious policy on the part of government! Both elements of this observation are borne out by the persistence of marked and predictable disparities in physician supply over space in most countries. However, some developing countries such as China and India have made service for a specified period in underdoctored or rural areas compulsory for some new medical students. In the USSR, there should be no disparities in physician supply if the system works efficiently but, as seen in Chapter 2, many inequalities do remain!

Areal variation in physician availability

Physician availability can be considered in two ways: availability as total supply and availability as geographical distribution (Joroff and Navarro 1971). The emphasis here is on the latter, but it is unwise to dismiss the question of total supply without giving some thought to its implications for access. Total supply is taken here to refer to the total number of physicians available to a health care system and is usually expressed in terms of population:doctor ratios (see Chapter 2).

Increasing the total supply of physicians has often been viewed as a means of tackling the problem of *regional* underservicing (Reilly et al. 1980). For example, the output of New Zealand medical schools was increased by 40 percent between 1968 and 1975 (Barnett and Sheerin 1978). Underlying this increase was the proposition that market saturation would eventually cause doctors to avoid overserviced areas. However, the achievement of this sort of intent depends first upon whether the increase in supply is sufficient to saturate attractive areas (assuming that overdoctoring is a symptom of attraction) and secondly upon the ability of physicians to circumvent the oversupply problem by creating additional units of demand for their services. The evidence to date suggests that supply policies such as those enacted in New Zealand (and similarly in Canada) are at best only partially successful (De Vise 1973; Barnett and Sheerin 1978; Gray 1980). In Canada, although the absolute number of physicians in most underserviced areas has increased since the mid-1960s, overserviced areas appear actually to have improved their *relative* positions (Joseph and Bantock 1982).

Despite the plausibility of this link between total supply and regional distribution, research on it remains handicapped by the difficulty of estimating the need for medical services and, by extension, the total number of physicians required in the system (see Chapter 3 for a fuller discussion of 'need' concepts). For example, Joroff and Navarro (1971) report that at various times between 1933 and 1959 authoritative estimates of the number of physicians needed in the United States ranged from 118 per 100,000 to 165 per 100,000, implying substantially different needs in terms of total supply. In Hong Kong, where physician location in public housing developments is planned very strictly according to specified norms, figures of 1 doctor per 6,000 persons have been altered to 1 per 10,000, which indicates very different *total* numbers required (Phillips 1983a). Research on regional distributions circumvents the problem of defining need by focusing on relative disparities in supply rather than on absolute numbers of physicians. Therefore, references to 'overdoctored' or 'underdoctored' areas are usually couched in relative, not absolute, terms.

Studies of regional disparities in physician availability can be categorized in a number of ways. In Chapter 5, for example, we shall classify research in terms of the structure of measures of access. Here, however, we shall distinguish between studies in terms of their treatment of space, namely macro-studies versus micro-studies.

Macro-studies of physician availability

Macro-studies have been based upon one or other of two mutually exclusive views of space. In one body of work, space has been viewed as a series of discrete spatial units or regions. Most often these regions have a political/administrative basis, such as states in the United States. Another body of work has viewed space in terms of its attributes, differentiating perhaps between 'rich' and 'poor' areas or, as is most common in the health care delivery literature, between urban and rural areas.

The persistence of the regional perspective on manpower distribution reflects the importance of administrative structures within health care delivery systems. In federal systems, such as those of Australia, Canada, and the United States, regional variations are virtually guaranteed through regional autonomy (or partial autonomy) in health care delivery. In Canada, for example, user fees for services vary between provinces as do schedules of payments for physicians. Even within non-federal systems, as in the United Kingdom and New Zealand, the regionalization of authorities over specific components of the health care system (for example, regional and district health authorities and, previously, regional hospital boards in the United Kingdom) provides a basis for a regional perspective.

Earlier in this chapter, we introduced the British system of characterizing physician availability patterns in terms of list size. Table 4.5 provides data for two dates on the distribution of relatively oversubscribed general practitioners (in terms of list size) by major regions in England. These data illustrate two important points. First, the range amongst the regions is considerable, both in 1966 and 1976. Second, despite the general improvement in the supply situation between 1966 and 1976, the relative position of regions hardly changed, the exception being the West Midlands, which jumped four rank positions. Interpretations of tabulations such as this are

Table 4.5 *Regional variations in primary health care provision in England: percentage of GPs with list sizes of more than 2,500 persons*

	1966		1976	
Standard Region	%	(Rank)	%	(Rank)
North	51	(4)	47	(6)
Yorkshire & Humberside	53	(5)	44	(5)
East Midlands	59	(7)	49	(8)
East Anglia	37	(2)	29	(2)
South East	47	(3)	37	(3)
South West	29	(1)	24	(1)
West Midlands	61	(8)	43	(4)
North West	55	(6)	48	(7)

Source: Phillips (1981a)

sometimes couched in terms of the broad socio-physical characteristics of the regions – for example, 'the South West provides attractive physical and social environments for physicians', – but are of limited usefulness because of the difficulty of generalizing about conditions within such large regions. This leads us to the alternative perspective that has been adopted in macro-studies: the attribute approach.

In the United Kingdom, the attribute approach has become intimately associated with the *inverse care law*. This term was coined by Hart and we can do no better than to repeat his seminal statement.

> In areas with most sickness and death, general practitioners have more work, larger lists, less hospital support, and inherit more clinically ineffective traditions of consultation than in the healthiest areas; and hospital doctors shoulder heavier case-loads with less staff and equipment, more obsolete buildings, and suffer recurrent crises in the availability of beds and replacement staff. These trends can be summed up as the inverse care law: that the availability of good medical care tends to vary inversely with the need of the population served.
> (Hart 1971, p. 412)

This statement in notable for a number of reasons. First and foremost, it does not specify a particular geographical scale. Hart's law can be taken to encompass variations within small, intraurban regions; for instance, the work of Salmond (1974) concerns access to perinatal care within the Wellington area, New Zealand, and Knox (1978) discusses differences in four Scottish cities. It can also apply across nations, and Bosanquet (1971) has suggested the river Trent as a boundary between the service-rich south of England and the service-poor North.

Second, it reintroduces *need* into the consideration of disparities in supply, where need is expressed in terms of morbidity and mortality rates. The implication is that local health status will increase if an improvement in health care occurs. Although the work of Hollingsworth (1981) and others (see Chapters 3 and 6) suggests that the relationship between health status and health care availability may be substantially modified by class-based health behaviour, it is hard to reject totally such a proposition.

Last, but not least, Hart's words draw attention to a sometimes neglected facet of the availability question: that of the quality of care versus its quantity. All too often researchers are seduced by the analytical simplicity that results from assuming that units of health care delivery – for instance, a general practitioner – are identical. This is rarely true (Knox 1979a), and this matter is discussed further in Chapter 8.

Dental care as well as medical care can be seen to vary markedly within regions. In the northern region of England, for example, the supply of dentists is very uneven and the region as a whole has some of the poorest levels of care nationally (Table 4.6). Within it, only Newcastle is better supplied than the national average and South Tyneside is in the unenviable position of having the worst ratio of patients to dentists in an urban area of England. Carmichael (1983) has also been able to show that many of these poorly provided divisions of this region have very poor dental health, which suggests the existence at this scale of an inverse care law.

Table 4.6   *Population/dentist ratio in the northern region of England*

Division of northern region	Persons per dentist (1979)	Difference from national average
Newcastle upon Tyne	3418	−419
North Tyneside	3860	+23
Cumbria	4233	+396
Northumberland	5163	+1326
Gateshead	5738	+1901
Cleveland	6114	+2277
Durham	6348	+2511
Sunderland	6400	+2563
South Tyneside	7391	+3554
Northern region	5187	+1350
England and Wales	3837	0
South Western region*	3225	−612

*included as an example of a relatively well-provided region

Source: after Carmichael (1983)

In the United States and to some extent in Australia, Canada, and New Zealand, macro-scale studies of physician availability have focused upon rurality as a critical attribute for the identification of geographical disparities in the availability of physicians. However, even in countries such as Great Britain, where the nature of rurality is different from that in the less densely populated countries of the 'Anglo New World', access to essential services such as health care has been identified repeatedly as a major rural problem (Moseley 1979; Phillips and Williams 1984). North American examples may be used to illustrate the major dimensions of rural-urban disparities in physician distribution.

> Rural America lags behind the rest of the country in both health status and the ability to compete with more urban localities for health resources.
> (Reilly et al. 1980, p. 120)

The existence of a chronic undersupply of physicians in rural America was brought to general attention by Mott and Roemer (1948), although the problem can undoubtedly be traced back to the second half of the nineteenth century.

In the largely preindustrial, preurban United States of 1850, the state of medical science and technology was such that the practice of medicine was not compromised by a rural location and, taken together with the rural upbringing of most physicians, there would appear to have been little inducement for urban location (Brown 1974). As the nineteenth century advanced, however, developments in the nature of medicine and the organizational and spatial infrastructures of delivery served to emphasize urban centres as the loci of medical care (Brown 1974; Knox and Bohland 1983). By 1914, about 50 percent of new medical graduates opened their practices in places of 100,000 people or more, whereas these places had only 26 percent of the nation's population (Roemer 1948). Such tendencies could not avoid creating massive disparities in the rural–urban availability of physicians. In 1906, rural America had 53 percent of the national population and 41 percent of all physicians. By 1926, the corresponding figures were 48 percent and 31 percent (Shannon and Dever 1974). Throughout the course of this century, the increasing specialization of medical manpower, coupled with rural depopulation, has served to guarantee a continuation of rural–urban disparities in physician distribution. This translates into a very real access problem, especially to higher levels of care. Because of the inability of small rural settlements to attract (or to retain) physicians (Fahs and Peterson 1965), patients are frequently forced to travel considerable distances for care (Kane 1969), often beyond perceived reasonable travel limits (Shannon et al. 1979). An almost identical situation has been noted in Australia (Brownlea and Ward 1976), in the USSR, and in a number of developing countries such as Sierra Leone (Hardiman and Midgley 1982).

The picture is not, however, quite as simple as orginally painted, chiefly because the nature of urbanization has changed during the course of this century. Beginning in the early 1950s, metropolitan and urban regions have displayed an increasing tendency toward diffusion or 'sprawl'. This resulted first in the growth of suburban nodes and more recently in the expansion of the non-farm population of rural areas proximate to urban population centres (Joseph and Smit 1981). Evidence from the American mid-west (Brown 1974) and from Alberta, Canada (Northcott 1980) strongly suggests that for general practitioners and specialists with relatively low levels of threshold demand (for example, general surgeons) this restructuring of metropolitan population distribution has resulted in a reduction of urban–rural disparities. There is, however, no evidence of any real convergence in manpower availability.

The major limitation of these urban–rural studies, and of macro-studies in general, is that an aggregate view may mask important variations *within* large regions or locality types. In connection with rural underservicing and to obtain a truly adequate appreciation of the nature and implications of manpower shortages, it is necessary to examine evidence on locational patterns *within* rural areas.

Micro-studies of physician availability

The majority of micro-studies of physician availability have examined distributions within *urban* areas. Several themes or approaches are evident in this work, ranging from central place interpretations of location patterns to attempts simultaneously to consider the demand and supply features of service delivery. The central place approach is an appropriate starting point, especially in the light of an almost universal tendency to consider broad regional changes within such a framework (Brown 1974; Shannon and Dever 1974).

Central Place Theory has often been applied to intraurban physician location, notably in the United States (De Vise 1973). The underlying proposition of this work is that physicians can be ordered in terms of the services they provide. Given a hierarchical ordering by specialty, one might expect the hierarchical frequency and locational arrangement of physician types to be determined by specialty threshold and range characteristics.

To date, one of the most extensive empirical investigations of central place regularities in physician location has been carried out in Phoenix, Arizona by Gober and Gordon (1980). The authors found that in 1970 there was an extremely uneven distribution of physicians of all specialty types in the Phoenix area; in central place terms, all specialties other than general practitioners exhibited a greater magnitude of concentration than expected. Gober and Gordon suggested this to be a combined function of physicians' hospital orientation and their avoidance of low socio-economic status areas, particularly those associated with ethnic or racial minorities. Like Shannon and Dever (1974), they argued that this demonstrated that the locational pattern of physicians reflects *their* needs rather than those of the client population. The weight of the evidence summarized earlier in this chapter (see 'The behavioural approach') would appear to bear out this contention.

Extensions of the distributional approach represented by Central Place Theory have sought to investigate the organizational attributes of the supply system and/or to characterize its ecological correlates. On the organizational side, Stimson (1981) has suggested a spatial competitive basis for the emergence of group practice as the dominant form of general practitioner organization in urban areas. Cleland et al. (1977a) noted that, in Adelaide, Australia, group practices were relatively more common in low-status and new-growth suburbs than in high-status and better-established areas of the city. Together with a higher population:physician ratio, this indicates a situation in which the areas with fewer physicians had them concentrated at a smaller number of locations. Double jeopardy!

In terms of the ecological correlates of supply, the evidence overwhelmingly supports Hart's (1971) 'inverse care law'. Cleland et al. (1977a) also found that access to (general practitioner) service opportunities in Adelaide was strongly and positively correlated with neighbourhood social status. Stimson (1980), again working with Adelaide data, showed disparities in potential access to be even more acute when access was considered in terms of surgery opening hours as well as spatial terms (Figure 4.3). Similarly, analysing data for several Scottish cities, Knox (1978) found that, with the exception of core areas which were well serviced regardless of class, access to general practitioner services appeared to be inversely correlated with neighbourhood socio-economic status. Moreover, Knox (1979a) has suggested that the apparently satisfactory access status of lower social status areas in doctor-rich core areas of cities may be illusory if the *quality* of service available is taken into account. Inner-city physicians may be older than average and work on their own out of dilapidated surgeries with very limited opening hours and little or no support from primary health care ancillary workers.

Earlier, with reference to the Phoenix study (Gober and Gordon 1980), it was suggested that ethnic or racial status might be an important ecological correlate of

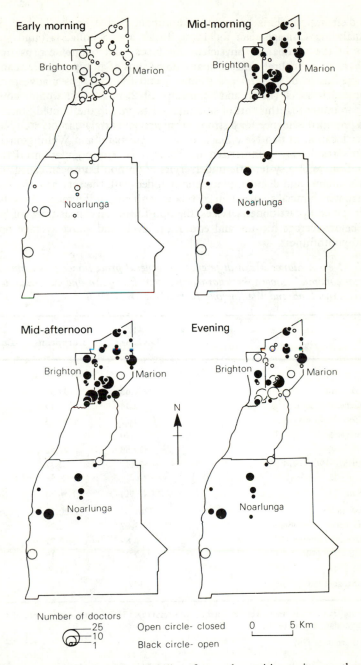

Figure 4.3 Spatial and temporal availability of general practitioners in a southern sector of Adelaide, 1977

Source: Stimson (1980)

physician distribution. This has been a dominant theme in the United States, of course. Analysing Chicago data for 1960, Elesh and Schollaert found substantial differences in the number of physicians in 'black' and 'white' census tracts: 'On average, there are over three times as many physicians of all types, more than two and one-half times as many general practitioners, and five times as many specialists in white as in black tracts' (Elesh and Schollaert 1972, p. 241). It would, however, be misleading to imply that this sort of situation exists only in the United States. Barnett (1978) has provided evidence for a strong ethnic/racial correlate in his study of general practitioner location in two New Zealand cities, Auckland and Wellington-Hutt.

Barnett's results are summarized in Table 4.7. The population:GP ratios are strongly indicative of a systematic underservicing on non-European census divisions. The socio-economic and demographic data included with the service ratios emphasize the degree to which racial status often correlates with class differences. As well as being deficient in general practitioner services, the non-European census tracts of both cities displayed below-average income and education levels and above-average population change and youthfulness.

Table 4.7 *New Zealand: Mean differences in general practitioner provision and selected socio-economic/demographic characteristics in European and Non-European census tracts, Auckland and Wellington-Hutt urban areas 1974*

Urban area	All Divisions	European	Non-European
Auckland	(N = 138)	(N = 107)	(N = 31)
Population:GP ratio	2054.68	1704.00	3756.88
Number of General Practitioners	2.44	3.19	1.50
Incomes $NZ6000 and over (%)	4.77	5.68	1.43
University education	4.70	5.48	1.89
Age 0–14 (%)	29.58	27.92	34.25
Population change 1966–1974 (%)	28.72	23.77	39.69
Wellington-Hutt	(N = 90)	(N = 76)	(N = 14)
Population:GP ratio	2261.38	1940.70	3879.20
Number of General Practitioners	1.87	1.99	0.96
Incomes $NZ6000 and over (%)	6.89	8.27	1.70
University education (%)	7.73	9.15	2.46
Age 0–14 (%)	28.55	27.99	32.49
Population change 1966–1974 (%)	13.81	10.19	36.28

Note: Non-European census tracts were defined as those having 15% or more of their population of Maori or Polynesian origin (1971).

Source: after Barnett (1978)

At various points in this chapter, we have referred to the increasing disparities in the urban–rural supply of physicians and resultant problems in rural health care provision. In describing the ecological approach to investigating physician location, for instance, evidence was introduced on the progressive development of rural–urban disparities in the United States. Such studies invariably take an aggregate view of space in order to incorporate effectively the temporal dimension. There is something of an empirical void between studies that examine broad changes in the rural supply of physicians (Reilly et al. 1980) and those that advance mostly anecdotal information on the changing location of physicians within rural areas (Bernstein et al. 1979). The work of Brown (1974) and Northcott (1980) reviewed earlier partly fills this niche, but both studies are limited by their time scope, as they concern 1950–1970 and 1956–1976 respectively. Consequently, this section is concluded by reference to a recent study by Joseph and Bantock (1983) which examines the locational pattern of general practitioners *within* a rural region of Ontario, Canada, through the course of this century.

The Joseph and Bantock study examines population and physician data for two counties in rural Ontario, Bruce and Grey, for 1901, 1921, 1941, 1961 and 1981. Equating demand for general practitioner services with population size, and supply with the number of physicians, the overall demand and supply position at each of the five selected time points is summarized in Table 4.8.

Table 4.8 *Physicians in Bruce and Grey Counties, Ontario: supply and potential demand over time*

	Population					
	Total	**Rural**		**Other**		
		No.	**(%)**	**No.**	**(%)**	
1901	127,748	95,406	(74.7)	32,342	(25.3)	
1921	102,456	63,972	(62.4)	38,484	(37.6)	
1941	98,025	55,856	(57.0)	42,169	(43.0)	
1961	102,896	48,597	(47.2)	54,299	(52.8)	
1981	130,794	62,963	(48.1)	67,831	(51.9)	

	Physicians					Total population per physician		
	Total	**General**		**Specialists**		**Total**	**General**	**Specialists**
		No.	**(%)**	**No.**	**(%)**			
1901	112	112	(100)			1,141	1,141	–
1921	87	87	(100)			1,178	1,178	–
1941	91	83	(91.2)	8	(8.8)	1,077	1,181	12,253
1961	80	62	(77.5)	18	(22.5)	1,286	1,660	5,716
1981	159	108	(67.9)	51	(32.1)	823	1,211	2,565

Note: **Other** refers to the population of incorporated places. The number of incorporated places increased from 17 to 25 from 1901 to 1921 and remained stable at 26 for the remainder of the study years. The minimum population size of incorporated places ranged between 300 and 500 over the study period, while their mean size fluctuated between 1,500 and 2,600.

Source: Joseph and Bantock (1983)

Several aspects of these data are worthy of comment. In connection with population (potential demand for services), the progressive depopulation of rural regions characteristic of most of central Canada during the first part of this century is clearly evident for Bruce and Grey counties. The resurgence in population after 1961 is largely attributable to the expansion of metropolitan influence out from urban core areas (Joseph and Smit 1981), which in this region has been most evident in terms of recreation- and retirement-related development emanating from Toronto, London, and other nearby urban centres. Additionally, the progressive absolute and relative concentration of the study area population into incorporated places is notable.

Table 4.9 *Bruce and Grey Counties, Ontario: location of physicians by settlement type*

| | **All Physicians** | | | |
	No.	Incorporated (%)	No.	Unincorporated (%)
1901	70	(62.5)	42	(37.5)
1921	74	(85.0)	13	(15.0)
1941	87	(96.0)	4	(4.0)
1961	78	(97.5)	2	(2.5)
1981	151	(95.0)	8	(5.0)

| | **General Practitioners** | | | |
	No.	Incorporated (%)	No.	Unincorporated (%)
1901	70	(62.5)	42	(37.5)
1921	74	(85.0)	13	(15.0)
1941	79	(95.0)	4	(5.0)
1961	60	(9.70)	2	(3.0)
1981	101	(94.0)	7	(6.0)

| | **Specialists** | | | |
	No.	Incorporated (%)	No.	Unincorporated (%)
1901	–		–	
1921	–		–	
1941	8	(100)	–	
1961	18	(100)	–	
1981	50	(98.0)	1	(2.0)

Source: Joseph and Bantock (1983)

In terms of physician supply, two things are evident from Table 4.8. First, the decline in total numbers of physicians parallels that of population in the first half of the century but continues through to 1961 despite the upswing in the latter. This is reflected in the peaking of the population:physician ratio in 1961. As noted earlier in this chapter, there is evidence that this 'shortage' of physicians was fairly widespread in non-metropolitan Ontario in the early 1960s (Bean 1967; Kerr 1967; Spaulding and Spitzer 1972). Similarly, the substantial increase in the number of physicians in the two counties between 1961 and 1981 is at least partly attributable to provincewide developments. The other major feature of the physician (supply) data is the increase in specialization. It is notable, however, that the levels of specialization in the (*rural*) study area were considerably lower than for the province as a whole, which were 16.7 percent in 1941, 39.5 percent in 1961, and 44.9 percent in 1981. Moreover, Gray (1980) suggests that these provincial figures may themselves be underestimates.

What then of the location of physicians *within* the two counties? The data on physician location indicate a radical restructuring of the general practitioner network between 1901 and 1941 (Table 4.9 and Figure 4.4). In that period, and especially before 1921, the general practitioner virtually disappeared from the unincorporated places of the two counties. In more specific terms, this meant a consolidation of general practitioner services into larger villages and towns (which were generally 'incorporated' as independent municipalities) and away from hamlets and small villages. Joseph and Bantock (1983) suggest that this restructuring of the service network between 1901 and 1981 reflects a fluid interaction between the changing distribution and mobility of the rural population and the organizational norms of health care delivery. Notwithstanding the precise mechanisms involved, the consolidation of general practitioners into larger, incorporated places resulted in the widening of disparities in access within the region. In spite of increases in mobility, the authors conclude that concentration of service has left the dispersed rural population at a lasting disadvantage relative to their urban counterparts and to ruralites fortunate enough to live in an incorporated place with a physician.

Summary

The evidence presented in the previous section suggests that disparities in the supply of (primary) physicians relative to potential demand for services is marked at all geographical scales. Moreover, disparities appear to be related systematically to the locational attitudes and preferences of physicians and to the properties of the systems in which they work. However, it is not always easy to distinguish the impacts of attitudes and preferences from those of system constraints. For example, underservicing of a rural community might reflect a lack of medical infrastructure (hospitals), a negative lifestyle image on the part of physicians, or some combination of these and other factors. The net result is that attempts to manipulate the physician location system based only upon individual aspects of physician preference (e.g. financial incentive schemes) or on individual constraints (e.g. health care clinics) have met with partial success at best. Experience to date suggests that, in the absence of complete government control of physician location, success may depend on a wide-spectrum approach to the problem, addressing perhaps physician attitudes, financial reward mechanisms, infrastructure development, and so on, at the same time.

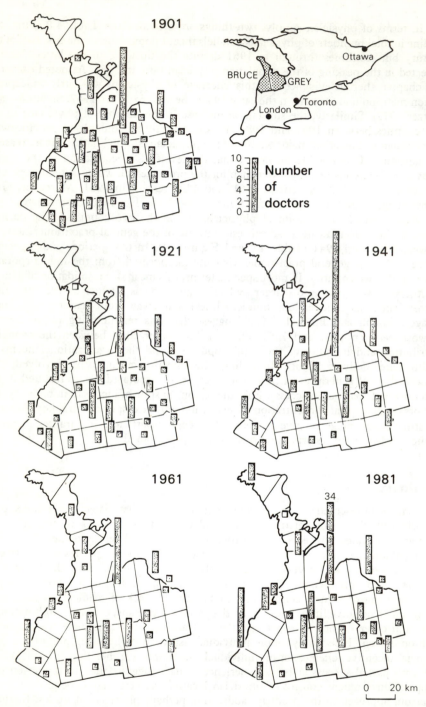

Figure 4.4 The changing location of general practitioners in Bruce and Grey Counties, Ontario, 1901–1981

Source: Joseph and Bantock (1983)

It is apt to conclude this chapter with a reiteration of an earlier point concerning the mutually reinforcing relationship between pattern and process. The majority of the developments in medical science and technology that have unfolded at an increasing pace since the start of this century have necessitated increased specialization of manpower and facilities at least in the *developed* world. However, the *specific* organizational and locational structure of health care delivery systems has been shaped by other forces, amongst which the locational dynamics of physicians must rank highly. It appears, however, that existing patterns of disparity become enshrined in contemporary attitudes. Place X is a bad place to practise because nobody practises there!

5 Measuring the Potential Physical Accessibility of General Practitioner Services

The evidence for regional disparities in the availability of physicians summarized in the latter parts of the previous chapter emphasizes the importance of measuring the various facets of 'performance' in a health care delivery system. In the past it has been common to focus monitoring efforts on utilization rates or on aggregate health outcomes. Planners and decision-makers might, for instance, use hospital admission rates or regional mortality rates to gauge the changing access to health care of a specific group within a client population. However, as discussed in Chapter 3, these measures are difficult to use.

Although measures of utilization or outcome may reflect changes in the health care system, only in the exceptional circumstance when a single element in the health care delivery picture changes over a given time period (all others remaining constant) can changes in utilization or outcome be unambiguously attributed to specific developments. An observed decline over time in infant mortality amongst a specified subgroup of a population might result, for instance, from improvements in medical technology, cultural changes in childbearing (for example, an increasing proportion of younger mothers), greater availability of health care, higher standards of living, or some combination of these and other, perhaps unmeasurable, factors. Rarely, if ever, does the student of health care have the opportunity to observe the impact of individual changes in isolation. Consequently, for purposes of monitoring individual aspects of health care delivery, specific measures of potential accessibility are easier to put into operation and interpret than measures of revealed accessibility. Potential measures include those focusing on the socio-economic and organizational dimensions of health care delivery, but the concern here is with measures of potential physical (or geographical) accessibility. The task of such measures is to assess the nature and pattern over space of physical access to health care, and to permit evaluation of the implications for physical accessibility of changes in the form and location of health care delivery.

Given the importance of the general practitioner within the health care delivery system (see Chapters 3 and 4), the design of measures of potential physical accessibility to *general practitioners* and a selective review of measures developed to date will provide the central focus for the discussion of the potential physical accessibility approach.

The design of potential physical accessibility measures

All measures of potential accessibility to general practitioners share a common task: to characterize variation across space in the potential accessibility of general practitioner services to individuals or groups. They also share a common need for data on demand (for general practitioner services) and supply (of general practitioner services over space). These data may be manipulated to produce a measure of access using two basic options: a 'regional availability' approach and a 'regional accessibility' approach.

The regional availability approach is the simpler of the two and involves an examination of the regional distribution of supply versus demand. Taking the United States as an example and equating the supply of general practitioner services with the number of general practitioners and the demand for their services with total population, one could calculate the ratio of population to general practitioners for each of the states. Furthermore, there is the possibility of subdividing states into counties and counties into municipalities. However, the calculation of availability ratios does have hazards at these increasing levels of spatial disaggregation, which can relate to the accuracy of data and uniqueness of places.

The regional availability approach carries the assumption that boundaries are impermeable, that is, that the population of any given region has access only to the general practitioner services available in that region. In some instances this assumption is quite tenable. In Canada, for example, the provincial basis for health care insurance makes it difficult for a resident of, say, Alberta to use regularly the services of a physician in neighbouring Saskatchewan. There are of course exceptions in the case of medical emergencies, students studying out-of-province, and so on. The county boundaries within provinces, by contrast, are totally permeable in this context. In general, the impermeability assumption is met only if regional boundaries correspond with an important organizational attribute of the health care delivery system, such as an insurance scheme or autonomous administrative unit.

The permeability problem is of limited importance if regions are large. In the example of the United Kingdom, given that it is known that most people use the services of a general practitioner in their local area, it is reasonable to assume that the vast majority of the inhabitants of any county use physicians in that same county and, conversely, that an overwhelming proportion of physician service in any county is used 'locally'. However, as the level of aggregation decreases, and particularly in border zones, the probability that individuals might use the services of physicians in a district other than their own increases. At very low levels of spatial aggregation, such as the urban census tract, this permeability problem would seriously compromise the regional availability approach unless the flows in each direction across regional boundaries were always exactly counterbalancing.

The second approach to the manipulation of demand and supply data, the regional accessibility method, acknowledges and accommodates the potential for complex interaction between supply and demand located in different regions. Measures using this approach are usually based upon spatial interaction (gravity model) formulations. As such, they are more complex than the ratios commonly associated with the regional availability approach and more demanding in their inputs, in that the structure of the distance decay function as well as the location of supply and demand must be specified.

A major limitation of the regional accessibility approach stems from the need to allocate regional demand and supply totals to discrete points, such as regional centroids, to permit the calculation of interregional distances (Shannon et al. 1969). The accessibility of a region is then equated with that of its centroid. This assumption has little impact on the accuracy of results when regions are small but, as regions increase in size, it becomes progressively less reasonable to imply that the mutual accessibility of two points at the boundary of two regions and separated by only a few miles would be the same as that of their regional centroids, separated perhaps by tens or hundreds of miles. It would, for instance, be patently unreasonable to use a regional accessibility measure at the state level in the United States or the provincial level in Canada.

In summary, the two approaches to the manipulation of supply and demand data are not equally sensitive to the scale of analysis. If levels of spatial aggregation are high, and even if boundaries are permeable, regional availability measures are more suitable than regional accessibility measures but, at low levels of aggregation, regional accessibility measures hold a clear advantage.

Clearly, there is a 'grey area' in terms of choice of approach, in which the strengths and weaknesses of each appear to trade off almost perfectly. It is impossible to attach distance values to this 'grey area' because it will vary over space and time, even within the same national health care delivery system. However, the researcher may gain some guidance from evidence produced in studies of general practitioner utilization discussed in Chapter 6.

The above classification of approaches to measuring potential physical accessibility is very simple but useful nevertheless as a springboard to further discussion. The remainder of this chapter is devoted to a more detailed consideration of the major attributes and relative merits of representative measures of both the regional availability and regional accessibility type.

Regional availability measures

Studies of the regional availability of general practitioner services and other elements of the health care system relative to potential demand have proliferated in the last decade or so. This has occurred as social scientists have become increasingly aware of the potential for the inequitable allocation of resources, inherently the subject-matter of the 'welfare approach' in human geography (Smith 1974, 1977). The various measures of the regional availability of general practitioners relative to potential demand that have appeared in the literature can be distinguished on three bases: level of spatial aggregation, data inputs, and data manipulation. In principle, considerable variety is possible in the specification of measures in relation to all three of these criteria but, in practice, data inputs and data manipulation have varied little amongst measures, particularly in contrast to level of spatial aggregation. This can best be demonstrated through a review of some representative works incorporating measures of the regional availability of general practitioners.

For example, at one extreme, a high level of spatial aggregation is adopted by Roos et al. (1976), who examine population to physician ratios in the Canadian provinces. Only slightly less spatially disaggregated is the work of Spaulding and Spitzer (1972), who consider population per primary physician contrasts between the northern and southern segments of Ontario. Both of these works also consider variation at other scales: rural–urban subregions in the case of Roos et al. (1976), and rural–urban and urban–urban contrasts in the case of Spaulding and Spitzer (1972). Examples of availability measures applied *within* urban areas are provided by Barnett (1978) for the Auckland and Wellington-Hutt metropolitan areas and by Stimson (1981) for Adelaide.

Data inputs

In contrast to the variety of scales adopted (some of the major implications of which will be discussed later), measures to date have on the whole used similar data inputs and data manipulations. Data inputs for measures of the regional availability of general practitioner services are of two types: data concerning supply and data on demand.

On the supply side, it has been common to equate the availability of general practitioner service in a region with the number of general practitioners in that same region. This assumption has at least two weaknesses. First, it is possible for general practitioners to be partly specialized. An example might be the rural general practitioner who doubles as a consulting obstetrician in the local community hospital. Part of this general practitioner 'unit' might be discounted, based perhaps upon his/her own estimation of time allocation or upon data derived from financial records (Barnett 1978). Stimson (1981) has gone further than most in overcoming this problem by using average number of consultations per hour and practice type (by 'group' size for instance) to weight physician availability, which is a useful refinement. Perhaps more important and far-reaching than the problem of specialization is that of physician quality. The more general problems of variation in quality of medical services are discussed in Chapter 8 but, briefly, physicians may vary in innate ability, attitude to patients, and so on. Moreover, quality of service may depend also upon work-load. Therefore, in an underdoctored area, potential clients may suffer not only through undersupply but also because this undersupply compromises the effectiveness of the service that is available. Such differences in quality are difficult to quantify and rarely, if ever, are they included in availability measures. Of course, variations in quality can cause considerable problems for health services planners, as discussed in Chapter 8.

On the demand side, the conventional practice has been to use total population as the numerator in the demand:supply ratio (Phillips 1981a). The use of total population as a surrogate for demand (or, more correctly, potential demand) stems from the difficulty of defining 'need' for health care (Boulding 1966; Bradshaw 1972), which was discussed in Chapter 3. However, it is pertinent to recall that there is strong evidence for the role of environmental, socio-economic, and cultural factors in determining health and ill-health, and thereby potential demand for care (Phillips 1979a; Fiedler 1981). Additionally, as developed in Chapter 6, there is overwhelming evidence for the U-shaped relationship between potential demand and age, such that populations with an above-average proportion of the very young and the very old will generate above-average demands for health care (Guzick 1978; Heenan 1980), and for the biologically determined proclivity of females to illness (MacMahon and Puch 1970). Consequently, it is not surprising that attempts have been made to disaggregate the population element of the demand:supply ratio. For example, in their study of the Vermont health care system, Evans and Chen (1977) calculated availability ratios for gynaecologists using the number of women aged between 15 and 44 as the population component. Similarly, they used numbers of children aged between 1 and 14 in their calculation of paediatrician ratios. Although this sort of disaggregation does not obviate the need definition problem, it is nevertheless a step in the right direction.

It is difficult to gauge precisely the potential impact of these data input problems on the results generated by measures of the regional availability of general practitioner services relative to population, but it seems obvious that they would increase with the level of spatial disaggregation. Differences between supply and demand units (for example, in age and sex structure) would tend to be smoothed out as the level of aggregation is increased. This decreasing variability with increasing spatial aggregation is an effect familiar to social scientists.

Distributional indices

In terms of data manipulation, it has been rare to come across anything beyond the tabulation and presentation of population:general practitioner ratios, perhaps with reference to some acknowledged norm or system average. The British practice of identifying areas with a list size of 2,500 or more as specially underserviced represents an excellent example of the use of service norms, whilst the Vermont study reported by Evans and Chen (1977) constitutes an equally typical example of the use of system averages. Most notably, little advantage has been taken of index measures designed to summarize regional distributions. Exceptions are provided by Stimson (1980), who uses an 'index of concentration' (a form of location quotient) to characterize variation in population:general practitioner ratios across the Local Government Areas of Adelaide, and by Knox (1979b), who calculates location quotients for the regional distributions of general practitioners versus population for Standard Regions in England and Health Board Regions in Scotland. Devices such as the *location quotient* facilitate easier and more accurate assessment of interregional differences at a given point in time and of the fate of specific regions relative to the system as a whole, over time.

Consider, for instance, a location quotient calculated as follows:

$$LQ_i^t = (GP_i^t/P_i^t) / (\sum_i GP_i^t / \sum_i P_i^t) \tag{1}$$

where:

LQ_i^t = location quotient for region i at time t;
GP_i^t = number of general practitioners in region i at time t;
P_i^t = population of region i at time t.

A location quotient greater than 1.0 means that a region has more than its 'fair' share of general practitioners relative to its share of total population. Conversely, a value less than 1.0 means that an area has less than its 'fair' share of general practitioners. A value of unity means that an area has exactly the number of general practitioners warranted by its share of total population. This location quotient could be calculated for specific age groups, such as the under-14s or the over-65s, or for other subgroups in the population.

The user of location quotients should always bear in mind, however, that comparisons over time of individual values (perhaps intended to monitor change in a particularly disadvantaged area) have to be interpreted in the light of the system average upon which individual LQ values are based. For example, a change in a specific LQ value from 0.9 to 1.1 between time t and time $t+1$ year would be interpreted as beneficial if the system average remained constant or improved. If, however, the system average deteriorated (that is, there were to be more people per general practitioner), the region's relative position would have improved, but its absolute situation might well have deteriorated!

There are specific idiosyncrasies of the location quotient. For example, values below unity are compressed in the range 0.0–1.0, whereas values above unity may range from 1.0 to infinity, so that a location quotient of 0.5 reflects an amount of underservicing equivalent to the overservicing implied by a value of 2.0. In spite of these, it is surprising that it has not been used very much in the health care delivery context. Even more surprising is that there are apparently few examples to date of the application of regional concentration measures to general practitioner data although the work of Brown (1974) and that of Northcott (1980) constitute notable exceptions.

Measures of regional concentration have a long history in geography, particularly in economic geography, and include the *coefficient of geographical association*, the *coefficient of redistribution*, and the *coefficient of localization* (Joseph 1982). Of these, the coefficient of localization is the most general and has the added advantage of being structurally related to the location quotient. Indeed, the location quotient and coefficient of localization together constitute a useful analytical package which could have application in health services planning.

The coefficient of localization is a type of gini coefficient and measures the concentration across regions of a phenomenon of interest relative to that of a base magnitude (Lloyd and Dicken 1968). In terms of examining the regional distribution of general practitioners relative to population, the coefficient of localization might be defined as follows:

$$CL^t = \tfrac{1}{2}\sum_i \left| \frac{GP_i^t}{\sum\limits_i GP_i^t} - \frac{P_i^t}{\sum\limits_i P_i^t} \right| \tag{2}$$

where:

CL^t = coefficient of localization at time t;
GP_i^t = number of general practitioners in region i at time t;
P_i^t = population of region i at time t.

A coefficient value of 0.0 indicates that general practitioners are distributed across regions in exactly the same proportions as population. Values between 0.0 and 1.0 reflect increasing levels of localization, although a value of 1.0 is unattainable because, given that physicians are people, it would be impossible to have a situation in which general practitioners were concentrated in one or more regions and people in one or more *other* regions (Joseph 1982).

The coefficient of localization would be a valuable aid in describing changes in the regional concentration of general practitioner services over time. Although the authors are not familiar with an application of the coefficient of localization to general practitioner data, applications in other areas demonstrate the utility of the approach. Moran and Nason (1981), for instance, use the coefficient of localization to monitor shifts over time in the regional concentration of the New Zealand dairy herd relative to the land base. Moreover, Northcott (1980) has demonstrated the utility of regional concentration measures per se in a temporal context through application of a gini coefficient to data on the distribution of general practitioners in Alberta, Canada, for five-year intervals between 1956 and 1976.

Northcott (1980) used a gini coefficient to assess trends in the rural–urban distribution of general practitioners:

$$G = 1 - \sum_{i=0}^{k-1} (Y_{i+1} - Y_i)(X_i + X_{i+1}) \tag{3}$$

where X is the cumulated proportion of the population over k regional size categories and Y is the cumulative proportion of general practitioners over k regional size categories. G varies between -1.0 and $+1.0$, with positive values indicating a disproportionate concentration of general practitioners in larger places and negative values indicating one in smaller places. A value of 0.0 indicates that the service distribution is identical to that of population.

Technical limitations

Like any index-type measures, the coefficient of localization, and the gini coefficient itself, have limitations. Isard stresses three which are of a technical nature:

> First, a change in the degree of fineness of area classification will generally cause a change in the coefficient... Second, the value of the coefficient... is relative; it describes a given distribution in terms of a base distribution and is only as good as the base is relevant. Third, the value of the coefficient... will tend to vary, depending on how broadly the non-base magnitude (e.g. industry sector, income class, and occupation group) is defined. (Isard 1960, p. 270)

These comments are, of course, almost identical to those made earlier in connection with data inputs for regional availability measures. In addition, Joseph (1982) has stressed an additional problem in interpreting the coefficient of localization.

Given that the coefficient of localization, like other indices, is statistically untestable, an empirically derived value has to be interpreted relative to the measure's limits. Joseph (1982) demonstrates that although the lower limit of the coefficient (0.0) is stable, the upper limit is highly sensitive to the number of regions in the system under study, the distribution of the base magnitude, and the region(s) in which concentration is occurring. Consider, as an illustration, the concentration of general practitioners across the 10 Canadian provinces. Ontario, the most populous of the provinces, has approximately 36 percent of Canada's population, and Prince Edward Island, the least populated, has approximately 0.5 percent. If all Canada's general practitioners were to locate in Ontario, application of Equation (2) would produce a coefficient of localization of 0.64, whereas if they crowded on to Prince Edward Island the coefficient would be .995! The lesson to be learnt from this simple example is straightforward: the coefficient of localization is a measure of concentration *relative* to a base magnitude and not a measure of *absolute* concentration (Joseph 1982).

Similarly, the gini coefficient used by Northcott (1980) produces results which are subject to the limitations outlined by Isard (1960) and Joseph (1982). The gini coefficient is particularly sensitive to the researcher's design decisions in that regional size classes (number of classes and class boundaries) as well as areal units have to be defined. Indeed, given that the number of classes used is usually small, upper and lower limits rarely approach $+1.0$ and -1.0.

Scale limitations

Notwithstanding operational limitations arising from the specification of data inputs and their manipulation, the major factor limiting the potential usefulness of regional availability measures stems from the scale problem referred to in the opening section of this chapter. Given its design characteristics, it may well be impossible completely to overcome the scale problems of the regional availability approach.

At a high level of aggregation, a regional availability measure might well mask important subregional variations (Shannon et al. 1969). At the provincial level in Canada, for instance, ratios of people per general practitioner would mask substantial and important intraprovincial variations, for example, between the metropolitan core regions of southern Ontario and the peripheral, resource-based regions of northern Ontario (Spaulding and Spitzer 1972). Even at less aggregate levels, regional availability measures are likely to be too crude to detect and monitor the local significance and impact of changes in the organization of general practitioner services.

The most important development in the delivery of general practitioner services over the last 20 years or so has undoubtedly been the rise of the group practice and the concomitant decline of the solo GP (Chapter 4). However, because this trend has not been geographically uniform, the consequent reduction in overall accessibility *within* regions has not been even. Cleland et al. (1977b) provide evidence of developments in general practice for Adelaide, and Phillips (1981a) for West Glamorgan (Figure 4.2). Unfortunately, there is no way of reflecting differences in a measure of regional availability; the locational organization of general practitioners within a region is not taken into account in the calculation of a population:general practitioner ratio. In one region, 10 general practitioners might all work from one large health centre, while in another (of equal population and area) 10 general practitioners might work out of individual, spatially dispersed offices/surgeries. The accessibility of GPs to potential customers in these two different types of organization would probably vary considerably.

The problem of intraregional variability is obviously ameliorated when less aggregated spatial scales of analysis are used. However, this gain has to be traded off against the increasing problem of permeability (cross-boundary utilization flows), as outlined earlier in this chapter. This is exemplified in the work of Spaulding and Spitzer (1972).

As part of an analysis of medical manpower trends in Ontario during the 1960s, Spaulding and Spitzer (1972) compared the population:primary physician ratios (PPPR) of urban and rural areas. In a number of instances, they found the PPPR to be much higher in the rural parts of regions in comparison to urban cores. In the case of Kingston, a city of about 60,000 persons in eastern Ontario, the PPPR in 1971 was 1,040:1 whereas in the remainder of Frontenac Country (within which Kingston is located) the same ratio was in excess of 4,300:1. There might well have been a substantial difference in the potential physical accessibility of general practitioner services in Kingston and Frontenac in 1971 but it is unlikely that it was as marked as suggested by the PPPR ratios, mainly because inhabitants of Frontenac County located near to Kingston had access to general practitioners located there. Indeed, evidence for the United States suggests that in low population density regions, similar to eastern Ontario, general practitioners within 10 miles, or even more, should be considered 'accessible' (Ciocco and Altman 1954; Kane 1969). Similarly, evidence presented by Stimson (1980) for Adelaide suggests that trips in excess of 20 miles or more to visit a general practitioner are not uncommon within the fringe zones of metropolitan centres. (They may, of course, be undesirable from a medical and personal point of view.)

Two of the best examples to date of the application of availability measures at extremely low levels of spatial aggregation have been furnished by Barnett (1978), for Auckland and Wellington-Hutt, and by Stimson (1981), for Adelaide. Stimson's measure is the more complex of the two in terms of data inputs and its outputs (for instance, separate access scores are calculated for each of six time periods within a day) but is almost identical in basic design to the Barnett measure. The latter will therefore be used here to explore some of the issues that arise from the use of availability measures at the intraurban scale.

In each urban area Barnett's analysis was based upon urban census divisions: 138 in Auckland and 91 in Wellington-Hutt.

> For each census division estimates of the average population served per doctor... were derived by dividing the number of general practitioners whose surgeries were located within a one mile radius from the centre of the built-up area of a census division into the estimated... division population.
> (Barnett 1978, p.4)

Thus the Barnett measure departs from the traditional specification of population per physician ratios by defining the stock of physicians available to the population of a given area in terms of proximity to the centroid of that area rather than location within its boundaries. In this way it represents a combination of the traditional measure and the (spatial) opportunity index suggested by Wachs and Kumagai (1973).

The problems associated with the formulation of the Barnett (1978) measure have been discussed by Joseph (1981) and commented on, in reply, by Barnett (1981). First, individual surgeries might be located within a mile of the centre of the built-up area of more than one census division. In terms of traditional ratio measures of availability, this would constitute double-counting, in that individual doctors might be included in the denominator for more than one census division. Consequently, in comparison with the traditional ratio, Barnett's measure would produce different results for regions for which there were physicians located within one mile of the centroid of the region whereas it would produce identical results if there were no such physicians.

Second, even if the double-counting problem is set aside, the exclusion of physicians more than one mile from 'the centre of the built-up area' may be unrealistic given evidence of intraurban utilization patterns (Chapter 6). For example, in a study of utilization patterns in the Swansea region, Phillips (1979a) found considerable variation amongst his four study areas, with the percentage of individuals located more than one mile from their general practitioner's surgery ranging from 22.6 to 100.

The decision on which physicians are, or are not, accessible to the population of a given area is obviously critical in the application of measures like those proposed by Barnett (1978) and Stimson (1981). However, even if the range of general practitioner services within a particular regional system can be identified, it is clearly mistaken to imply, as do the Barnett and Stimson measures, that general practitioners located at the limit of a region's range are as accessible to its population as those located near to its centroid. This leads rather conveniently to a consideration of measures based upon the gravity concept, which can more readily accommodate known (spatial) regularities in the use of general practitioner services, such as the distance decay effect in consultation rates.

Measures of intraregional accessibility

Measures of regional accessibility based upon the gravity model are, given the long tradition of potential models in economic geography, a surprisingly recent feature in the literature on access to general practitioners. Early examples of the gravity approach are provided by the work of Guptill (1975) and Schultz (1975). More recent examples are Knox (1978) and Joseph and Bantock (1982). In comparison with regional availability measures, the major feature of regional accessibility measures is their ability to refine the treatment of demand and supply locational relationships through reference to utilization behaviour.

Distance decay

The primary intention of regional accessibility measures is to quantify the potential impact of demand and supply locational relationships on the utilization of general practitioner services. The relevance of evidence on actual utilization of general practitioner services is therefore obvious. Studies of utilization have consistently demonstrated that distance acts as a deterrent to therapeutic behaviour (Shannon and Dever 1974). This is evidenced by distance decay in consultation rates in environments as contrasting as rural Newfoundland (Girt 1973) and metropolitan Liverpool (Hopkins et al. 1968), some of which are discussed in Chapter 6.

The inclusion of distance decay as a central feature of a regional accessibility measure brings with it an important practical requirement: the specification of the distance decay function. Morrill and Kelley (1970) provide a succinct discussion of the specification of distance decay in the modelling of health care delivery. They suggest three major options: power functions of the gravity type (D^{-b}), negative exponential functions (e^{-bD}), and a hybrid of the two (for example, $D^{-b} e^{-bD}$). Each represents a different conceptualization of the potential impact of distance on utilization. With $b=2$, for instance, the exponential function implies a constant degree of deterioration in potential accessibility with each additional distance unit of separation from the source (1, 1/2, 1/4, 1/8, etc.) and the power function a constant but accelerating loss of accessibility with increasing separation (1, 1/4, 1/9, 1/10, etc.). The hybrid function, on the other hand, implies a decline of accessibility with increasing distance but at a decreasing rate (Morrill and Kelley 1970).

In the absence of any priori theoretical justification, the choice of the most appropriate distance decay function for use in a particular situation has frequently been determined by fitting functions to comparable data sets (Taylor 1977b). In the context of identifying the most appropriate distance decay function for use in a measure of the potential accessibility of GPs, this strategy is made difficult by a paucity of studies of general practitioner service utilization sufficiently detailed to provide enough data for calibration (see Chapter 6). Moreover, given the fact that distance decay effects will not be perfect because of the concurrent and complex interaction of spatial and non-spatial influences on utilization, it is clear that an extensive number of empirical studies of utilization is a prerequisite for identifying appropriate distance decay functions. In addition, these studies need to be carried out under diverse conditions in order to establish the sensitivity of distance decay to geographic and other structures (a tall order indeed!). Therefore, given the 'state of the art' in studies of utilization, there is no universally accepted distance decay function; and specification of its form in individual measures of potential accessibility is almost totally at the discretion of the researcher. However, given the variety of social and economic circumstances of users and the multiplicity of health care delivery systems, the lack of a common distance decay function is hardly surprising. Nevertheless, it must still be considered a major weakness (Barnett 1981).

Two measures

The two measures of the regional accessibility of general practitioners outlined by Knox (1978) and Joseph and Bantock (1982) serve as a useful focus for further discussion of the regional accessibility approach.

a) Knox developed his measure in the context of research into intraurban variability in access to general practitioner services in Britain. Applications of the measure are outlined in Knox (1978, 1979a, 1979b). The Knox measure begins with a calculation of nodal accessibility, in this case the nodes representing urban neighbourhoods.

$$A_i = \sum_{j=1}^{n} \frac{(S_j)}{D_{ij}^{\ k}} \tag{4}$$

where:

A_i = accessibility in neighbourhood i;
S_j = size of surgery facilities in neighbourhood j;
D_{ij} = linear distance between the geometric centres of neighbourhoods i and j;
k = distance decay function.

Knox measured S_j in terms of the total number of hours of consultation time available in a specific neighbourhood and the distance decay function chosen was a negative exponential one. The precise form of this function ($e^{-1.52D}$) was identified on the basis of a regression analysis of patient registration data presented in an earlier study by Hopkins et al. (1968).

The estimate of neighbourhood accessibility provided by Equation (4) was modified by incorporating measures of the mobility of residents of the various neighbourhoods, with (percentage) levels of car ownership in neighbourhoods (C_i) being taken as a surrogate measure of relative accessibility.

$$A_i(t) = C_i \frac{(A_i)}{T_c} + (100 - C_i) \frac{(A_i)}{T_p} \tag{5}$$

where $A_i(t)$ is the time-based index of accessibility for neighbourhood i and T_c and T_p are empirically derived estimates of the average time taken to travel a unit distance from the geometric centres of neighbourhoods by car and public transport, respectively. $A_i(t)$ is a measure of the potential accessibility of neighbourhoods (as areas). To account for differences in neighbourhood population, Knox weighted the index in terms of the population potential (M_i) of each neighbourhood,

$$M_i = \sum_{j=1}^{n} \frac{(P_j)}{D_{ij}^{\ k}} \tag{6}$$

where P_j is the population of neighbourhood j. Scaling $A_i(t)$ and M_i as a percentage of their respective highest computed values produces the final index of accessibility (I_i).

$$I_i = \frac{A_i(t)[\%]}{M_i[\%]} \tag{7}$$

Values exceeding 100 indicate a relative overprovision of physicians in a neighbourhood, whereas values below 100 indicate a relative underprovision.

The Knox measure is, on the whole, quite sophisticated, including, as it does, a time-based measure of the supply of general practitioner services (instead of just counting physicians) and a factor representing the differential mobility of neighbourhood populations. However, the measure does have certain limitations, two of which are acknowledged by Knox (1978) himself. First, there is the possibility for movement across the boundary of the system (that is, people outside the set of i neighbourhoods may use general practitioners within it and, similarly, individuals in one of the i neighbourhoods may use general practitioners located outside the i neighbourhoods). The potential impact of this boundary problem on estimates of accessibility is, of course, generally confined to neighbourhoods at the edge of the study area, and awareness of this should imbue the researcher with caution in the interpretation of unusually high or low estimates of accessibility for peripheral neighbourhoods. This problem can only be ignored if patients are actually assigned to specific physicians or service locations in a particular health care system (see Chapter 2).

Second, and more important, Knox acknowledges that his index measure provides only 'an *average* measure of accessibility, whereas in reality some important variations in accessibility will inevitably occur as a result of time-geographic constraints inherent to different sociospatial classes....' (Knox 1978, p. 426). Indeed, he also acknowledges that the measure could be made more sensitive by incorporating a finer mesh of neighbourhoods or through the inclusion of variables on age structure, morbidity, and mortality as indicators of comparative need.

In final comment on the Knox measure, two problems that affect all measures of potential physical accessibility can be noted. The first relates to the exponent of distance and the second to the elasticity of supply of physician services.

The problems associated with identifying the appropriate distance decay function for inclusion in a measure of potential physical accessibility have already been noted. In his measure, Knox (1978) uses a negative exponential function based on data derived from a study of *one* suburban Liverpool practice. However, the generality and transferability of the Liverpool result is problematic in the absence of other, comparable data. Moreover, there exists a more general and perplexing problem in connection with the distance decay component of accessibility models.

> The validity of including such (distance decay) exponents in gravity or potential type models is often confounded by their variability, that is, the extent to which the distance decay pattern of patient–doctor contacts is similar for different parts of the city. If the shape of the contact fields is significantly different between doctor-rich and doctor-poor areas, for example, then the inclusion of a general exponential value may be questionable. (Barnett 1981, p. 34)

This problem clearly applies to the Knox measure, although its precise impact is difficult to gauge, as it is also in the measure suggested earlier by Guptill (1975), because the inclusion of a mobility factor in Equation (5) serves to modify the general measure of accessibility in the light of conditions prevalent in individual neighbourhoods.

The second general problem, that of the elasticity of supply of physician services, is somewhat less perplexing. In the Knox formulation, individual S_j values (the measure of service supply) are available to all neighbourhoods, but they are modified by the degree of spatial separation between supply and demand zones. Given that the distribution of population and physicians is extremely unlikely to be anything like uniform (Knox 1978, 1979a, 1979b), the number of people within the catchment of general practitioners at different locations will vary considerably. Unless physician services are perfectly elastic in terms of demand, which they are extremely unlikely to be, the supply of services available to certain neighbourhoods will be underestimated, while that to others will be overestimated. Barnett (1978), for instance, notes a saturation effect in the use of general practitioners in the densely populated core areas of Auckland whilst other locations are presumably overprovided. In the light of Equation (7), the extent of overestimation or underestimation will depend upon the degree to which the average population in the catchment areas of individual physicians differs from the population potential (M_i) of each neighbourhood. The problem will be eliminated if population densities are uniform; and mollified if density gradients are not extreme. The issue of supply elasticity is therefore potentially less important in studies of intraurban accessibility than in studies of intraregional accessibility encompassing urban, rural and 'transition' areas.

b) By contrast with the Knox (1978) measure, that proposed by Joseph and Bantock was developed in an intraregional rather than an intraurban context. Like the Knox (1978) measure, it begins with a calculation of nodal accessibility,

$$A_i = \sum_j GP_j / d_{ij}^b \qquad (8)$$

where:
A_i = potential physical accessibility of rural enumeration area i to general practitioner services;
GP_j = general practitioners at j within the range of area i;
d_{ij} = distance between i and j;
b = exponent on distance.

As well as the obvious difference in the areal units for which accessibility is calculated (rural enumeration areas as opposed to urban neighbourhoods), this calculation of nodal accessibility differs from that of Knox (1978) in two ways. First, a limit is set on the distance ('range') within which a physician is considered accessible and, secondly, a power function rather than a negative exponential function is used to represent the distance decay effect.

After the initial calculation of nodal accessibility, the Knox (1978) and Joseph and Bantock (1982) measures part company. Rather than modifying nodal accessibility to reflect differential mobility, Joseph and Bantock chose instead to focus upon the impact of catchment populations on the potential availability of physician services.

$$D_j = \sum_j P_j / d_{ij}^b \qquad (9)$$

where the potential demand on a doctor at j (D_j) is a function of the magnitude of the population within the range of the service offered (that is, within the doctor's catchment area), modified by their distance away. Combination of Equations (8) and (9) produces (A_i^*) a measure of potential accessibility of general practitioner services to individuals incorporating a weighted estimate of physician availability.

$$A_i^* = \sum_j (GP_j / D_j) / d_{ij}^b \tag{10}$$

Joseph and Bantock evaluated their measure through a case study in Wellington County, southern Ontario, Canada. Physicians located outside the study area but known to take patients from the county were included in the analysis and the catchment population of these physicians was calculated using Equation (9). This circumvented the boundary problem mentioned earlier in connection with the Knox measure.

The accessibility index was calculated for three ranges of services values (5, 10 and 15 miles) but with the distance exponent fixed at -2.0. Again, it may be noted that in the absence of studies of utilization in comparable areas, researchers are left with only general indications of the appropriateness of a given distance decay function.

Calculated values for the Joseph and Bantock (1982) measure are mapped in Figure 5.1, in which scores have been raised by a constant of 10^4 to facilitate interpretation. Although the dimensionless nature of the measure makes it impossible to compare absolute enumeration area scores produced with different range of service values, changes in the relative pattern of potential physical accessibility are noticeable.

With a service range for GPs of 5 miles, a stark contrast exists between the potential physical accessibility of rural areas near urban centres (the location of the vast majority of general practitioners), and those less near (Figure 5.1a). Indeed, several enumeration areas are beyond the range of the nearest general practitioner and have a score of 0.0. When the service range is extended to 10 miles (Figure 5.1b), only one enumeration area has a potential physical accessibility score of 0.0 and, as would be expected, the overall pattern displays a systematic, though minor, reduction of the disparity between rural areas close to and far from urban centres. Extension of the service range to 15 miles has only a minor impact on the pattern of potential physical accessibility (Figure 5.1c). In fact, it appears that the measure is not very sensitive to change in the service range beyond that distance at which most, or all, areas are within the range of the service. This limited sensitivity stems from the progressive impact of the exponent on the distance in Equation (10), which exponentially reduces the contribution to potential accessibility of physicians at increasing distances.

As a final step in the analysis, potential physical accessibility scores were recalculated for a service range of 10 miles using Equation (8) instead of Equation (10) (Figure 5.1d). Comparison of the two patterns (Figures 5.1b and 5.1d) underlines the importance of weighting the availability of doctors, as represented by Equation (9). The unweighted pattern is dominated by the concentration of general practitioners in Guelph (52 of the 89 in the study): for instance, only one enumeration area in the four more northerly townships has a score above the lowest category. By contrast, the pattern produced by the weighted measure is more complex: in the same four northern townships, six enumeration areas have scores above the lowest category. By assuming each physician to be fully available regardless of the size of his catchment area population, the unweighted measure makes the disparity between rural areas near urban centres and those further away appear worse than it may actually be. Although physicians in more isolated rural areas may be some distance from potential clients, they are likely to be more 'available' to them, which may compensate in part for the inherent disadvantage of isolation from major, urban concentrations of general practitioners.

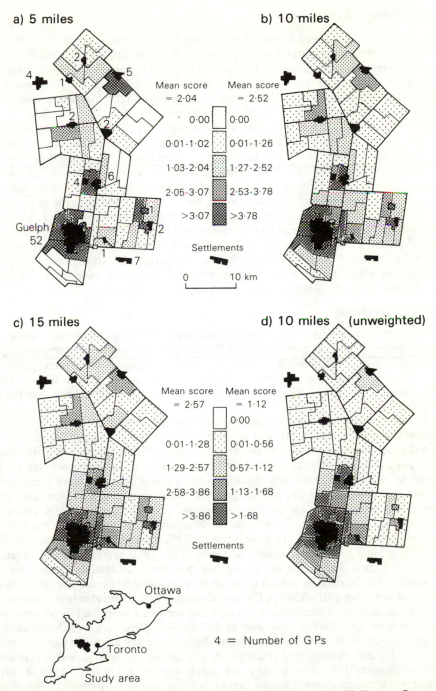

Figure 5.1 Potential physical accessibility to general practitioners in Wellington County, Southern Ontario (darker shading indicates greater accessibility)

Source: Joseph and Bantock (1982)

Technical limitations

The experiments performed by Joseph and Bantock (1982) underline the sensitivity of regional accessibility measures to the specification of exogenous (gravity model) inputs. Although measures of both the regional availability and regional accessibility type are sensitive to data inputs and the specification of the regional framework, regional accessibility measures are, in addition, highly responsive to the manner in which the distance decay component is designed and calibrated.

In principle (and, it may be hoped, in the medium term), the calibration problem could be tackled by improving knowledge of the behavioural underpinnings of health care utilization. However, in the short term, a more pragmatic approach may be necessary since it is unlikely that researchers will be able to abandon the use of accessibility measures. Although measures will still have to be calibrated in the light of available information, sketchy as that may be, every effort should be made to establish the sensitivity of results to these inherently suspect exogenous inputs. A possible procedure has already been outlined by Joseph and Bantock (1982) for their 'range of service' parameter. This sensitivity analysis, similar in nature and intent to constraint sensitivity analysis in linear programming, would permit the identification of those situations in which results are more responsive to changes in exogenously defined parameters than to variation in demand and supply conditions.

Future directions

It is abundantly clear from the previous discussion that there is considerable scope for improvements in both the regional availability and regional accessibility approaches to the measurement of the potential physical accessibility of general practitioner services. The measures currently available are reliable and sensitive enough to detect broad differences in availability or access but may not be a sufficient basis for policy formulation.

In terms of ease of interpretation (surely a prerequisite for policy use!), regional availability measures hold a distinct advantage over regional accessibility measures. Even the most obtuse politician should be able to grasp the significance of a difference in the population per physician ratio between two regions. Moreover, as noted in the discussion of regional availability measures, ratios could be made more meaningful through improving inputs (for example, on the potential *need* of the population in a region) and through more comprehensive analysis of regional patterns (for example, using location quotients, the coefficient of localization, or some other method of regional analysis). However, the basic scale limitation persists. Regional availability measures become increasingly suspect as the level of spatial aggregation is decreased. Indeed, it is most likely that the divergence between *local* availability and *overall* accessibility will be strongly related to the level of spatial aggregation. For instance, in a study of physician availability in Detroit, Guptill (1975) found that only 16 percent of the variation in his subcommunity accessibility scores for groupings of census tracts scores could be attributed to covariation in the supply of physicians by subcommunity.

It is at lower levels of spatial aggregation that questions of allocative justice may be most pressing and, as developed in Chapter 8, it is often at the local scale that planners' decisions assume relevance as they are implemented. However, 'It is of little consolation to the consumer to know that he is living in a well-provided region of the country if, within his local environs, facilities are inaccessible and/or of low quality'

(Phillips 1981a, pp. 66–67). Therefore, the most pressing methodological issues may well be those connected with the development of improved measures of regional accessibility. Although measures such as those outlined by Knox (1978) and Joseph and Bantock (1982) are relatively simple, they provide the basis for refinement in design and inputs (for example, the incorporation into a single measure of variables to reflect physician availability and quality, potential population needs and mobility, and so on). However, detailed refinement of potential physical accessibility measures would not be advisable if effected against a backdrop of partial ignorance concerning the behavioural regularities in the utilization of primary health care services. This, of course, has been a major proviso in all aggregate consumer research whether in health services or other settings (Thomas 1976).

Stimson (1980) and Phillips (1981a), among others, have made a plea for further research into the behavioural bases of health care utilization. The increasing sophistication of measures of one predisposing element in the health care utilization process – physical accessibility – demands concomitant increases in knowledge of the nature of utilization behaviour and the variables affecting this. Although there have been substantial increments in knowledge of broad regularities in the geographical dimensions of health care utilization (see Chapter 6), studies of the utilization of general practitioner services have been relatively few. In the absence of a substantial stock of such studies, it has been extremely difficult adequately to specify distance decay functions in existing models: Knox (1978), for example, calibrated a negative exponential function from data supplied for a single practice by Hopkins et al. (1968), while, in the absence of relevant empirical evidence, Joseph and Bantock (1982) set the distance exponent in a power function at -2.0. In this context, the prospect of 'disaggregating' measures to take into account differential potential need and mobility through the population is daunting! Indeed, it must be concluded that the possibility for increasing the sophistication of measures of potential physical accessibility is tied firmly to research on 'revealed accessibility', or consumer behaviour patterns.

In conclusion, it is pertinent to consider an important constraint on the use of measures of the potential physical accessibility of general practitioner services (or any sector of health care for that matter) of both the regional availability and regional accessibility sort – the availability of necessary data. The worth of any social indicator, health-related or otherwise, is a function of its operational practicality as much as its conceptual integrity. It is difficult, for instance, to monitor the unmeasured! Consequently, if the intention is to use physical access measures to monitor changes in the health care delivery system over time, they must be designed in the light of anticipated availability of time-series data. Therefore, even if better theory gleaned from behavioural studies of utilization allows for the construction of sophisticated and highly sensitive measures of physical access, their application will depend on data being available. A useful analogy would be the development of trains in the late 1960s, where technically feasible speeds could not be achieved because of inferior track. One can only hope the increasing availability of computer hardware and software will encourage an expansion of (social) data banks away from the rigidity of the national censuses. Ultimately, if potential access measures are to become a viable policy tool, there will be a need for comprehensive banks of data related to health care.

6   Utilization of Health Care Facilities: Revealed Accessibility?

The previous chapter focused on methodological questions relating to the measurement of the potential physical accessibility of medical services, particularly primary care. However, it must be recognized that even when optimal conditions exist in terms of service distribution and proximity, utilization may or may not occur. Even if utilization does occur, it may be variable in its efficacy (what the patient gains by attending) or in its frequency (how long and how often he or she will attend). Indeed, effective utilization at the 'correct' time and as frequently as required can be regarded as a major indicator of success in health care provision.

It is apparent that spatial factors are not the only ones influencing this outcome. Medical sociologists in particular have stressed aspatial, social and economic factors, whilst they and other social scientists have developed various types of models to characterize major influences on utilization. This chapter will consider some of these attempts to specify the variables influencing behaviour and to posit their effects, and will then proceed to examine inherently spatial influences, particularly aiming to answer the questions whether, how and why distance from source of care exerts an influence on utilization.

Medical sociologists have found it important to distinguish between health behaviour and illness behaviour (Mechanic 1962, 1968). Health behaviour is 'the activity undertaken by a person who believes himself or herself to be healthy for the purpose of preventing disease, while illness behaviour is the activity undertaken by a person who feels ill for the purpose of defining that illness and seeking relief from it' (Cockerham 1978 p.77). Lay persons commonly define illness in terms of the ability or non-ability to carry out activities and, therefore, there will tend to be much greater urgency and less voluntary choice in illness behaviour than in health behaviour. It is for this reason that the distinction is of relevance to the study of utilization. Some geographers have also distinguished between illness behaviour and therapeutic behaviour, which relates to the decision to seek medical care once it has been recognized that a state of illness exists (Girt 1973). It is such therapeutic behaviour that has generally been studied by geographers, who have, it seems, assumed (perhaps erroneously) that health and illness behaviour and aetiological factors causing disease show no spatial variation. In some environments, these may be correct assumptions but they should be empirically assessed wherever possible.

With the decision to seek medical care, the individual hopes that the consultation and treatment sought will restore his health or 'steady state with the environment', the ecological equilibrium with which health is usually equated (Howe and Phillips 1983). The effect of distance of the individual from care on illness and therapeutic behaviour is interesting, as it may well have an increasing and a reducing effect at the same time. Moreover, it is unlikely that these effects will be resolved into a tendency for the probability (or frequency) of consultation to decline smoothly with increasing distance (Girt 1973). The distance decay examples recounted in this chapter need to be appreciated in the light of this knowledge and in the context of the numerous variables which seem to influence utilization in a broad sense.

Models of health services utilization: specifying the variables

Within the social sciences, a group of studies has emerged with the common theme of modelling the key features of population characteristics, service characteristics, and other variables in an attempt to explain or predict utilization behaviour. These have usually been distinct from the geographical models introduced in Chapter 5 because of their interest in mainly aspatial variables. They can be presented in a chronological order, which is matched by a more or less increasing sophistication in methodology. The first four are discussed in some detail by Veeder (1975), who considers that they present a useful summary of the more conceptually advanced research on human and health services in the past 25 years. The fifth model is particularly useful from a 'systems analysis' point of view as it enables components of utilization and health services to be at least conceptually distinguished.

1   1959–1960 The Rosenstock model: psychological-motivational determinants of health service utilization.
2   1964–1966 The Suchman model: socio-cultural and environmental determinants.
3   1968 The Anderson model: family life-cycle determinants.
4   1972 The Gross model: behavioural components.
5   1974 The Aday and Andersen model: a framework for the study of access; the system as a modifier.

The Rosenstock model, developed in the late 1950s, stresses that the emotional rather than cognitive 'beliefs' of a person are crucial to understanding utilization. It has greatly influenced social-psychological explanations of the ways in which healthy people seek to avoid illness (health behaviour). It was further developed as the Health Belief Model (Rosenstock 1966; Becker 1974). A person will be likely to use health services if he believes himself susceptible to the disease in question, that it could have serious consequences for himself or his family, and that its course can be prevented or ameliorated by some action on his part. This action must not be more troublesome than the disease itself and a key concept is that, once the psychological state of readiness to use health services exists, a 'cue' may trigger action. Although certain barriers such as costs, distances, and inconvenient hours may act to reduce or prevent attendance, cues such as reminders from physicians or preventive mass-media announcements may provide an impetus for utilization (Rosenstock 1960, 1966). This model is interesting in its early suggestion that distance may act as a barrier to the receipt of care even if the psychological readiness and awareness of need to utilize exists. Indeed, early evidence of the importance of *physical* accessibility is provided for both polio vaccination (Rosenstock et al. 1959) and preventive health examinations (Borsky and Sagen 1959).

Suchman's model of the mid-1960s laid emphasis on the influence of social groupings and linkages on utilization. Critical determinants of utilization behaviour are matters such as the social network of family and friends within which an individual finds himself (McKinlay 1972; Booth and Babchuk 1972) and also the 'lay referral system', by which lay persons who are believed to be authoritative are approached before the professional physician. Therefore, the levels of health knowledge of kin and contacts will be important in influencing utilization, and it is noted that attitudes to illness and awareness of treatment vary considerably among cultural groups, low socio-economic status and minority groups tending to be more isolated and to have lower factual levels of disease and treatment knowledge. They may be fearful or sceptical of medical care and become dependent on lay advice. Today, this type of behaviour may be very important in developing countries and amongst ethnic minority groups in North America and Europe. It has also been argued that a 'cosmopolitan' social structure is more likely to be related to a 'scientific' orientation to health and medicine whilst a 'parochial' or traditional society is more likely to hold popular or folk beliefs (Suchman 1964, 1966).

Anderson (1968) proposed a model explaining utilization behaviour in terms of a sequence of conditions which will tend to regulate the volume of services used, and it is sometimes referred to as a life-cycle determinants model. These can be summarized as a set of factors which may *predispose* towards utilization: family composition (age, sex, size, and marital status); social structure (occupations, social class, education, ethnicity); and health beliefs, relating to attitudes to physicians, health care and disease. Certain other factors would *enable* the utilization of services. First, the family would have to have sufficient resources in terms of income, savings, insurance, and access to a regular source of care. Second, the community in which they lived would have to have the requisite resources, perhaps expressed in terms of a certain ratio of hospital beds or physicians to the local population. These factors would therefore, by their presence or absence, enable or hinder utilization. Finally, however, the family would require the stimulus of need.

This is a useful, commonsense model but it does not explicitly include accessibility as a factor influencing utilization – perhaps its main weakness. The 'enabling factors' imply a more general availability or absence of resources rather than an explicit proximity measure. Transport availability and mobility are also omitted, although they would be implicitly included in their associations with socio-economic status and income. In discussion, Anderson suggests that 'need' variables are perhaps the most powerful predictors of utilization and also that the balance between factors may vary between different levels of service (dentists, primary physicians, and hospitals, for instance). This suggestion is supported by McKinlay (1972), who considers that many results are service-specific, meaning that caution must be exercised in extrapolating findings beyond the service in which they are first observed.

Advancing in sophistication but perhaps losing in clarity, Gross (1972) proposed a regression model operating within a 'behavioural' framework. Accessibility variables are included as well as the predisposing, enabling, and need components highlighted earlier by Anderson (1968) but, as Gross (1972, p. 75) states, 'we need to know a lot more about the *relative* explanatory powers of behavioural or predisposing variables, the enabling variables (including financial and spatial-temporal accessibility measures) and health level indicators on the utilization of health services' before the model can be put into operation. The problems of multicollinearity, reverse causality, and data gathering beset the empiricist in these types of study, and a complex interlinking is evident with little to guide the researcher as to the real relative importance of the factors represented.

The equation

$$U = f\,[E;\;P;\;A;\;H;\;X\,] + \epsilon$$

seeks to explain the utilzation (U) of various services by an individual (see Figure 3.3).

E = *enabling factors* such as health insurance status, family size, occupation, education, and income.

P = *predisposing factors* such as attitudes of the individual towards health care, health services, and physicians; health behaviour and knowledge of the existence of various services.

A = *accessibility factors* such as distance and/or time of individual from facility, appointment delay, and waiting times; availability of medical and physician services and a regular source of care.

H = *perceived health level* of individual and/or his family.

X = individual and areawide exogenous variables (age, sex, family size, race, education, and location).

ϵ = residual error term.

This model certainly includes many recognized and potential influences on utilization and, as Gross explains, a number of discrete hypotheses are buried within this single functional relationship. Time-lags may operate (indicated in Figure 3.3 by dotted lines), although the model itself may be attempting to represent in a static-structural manner what is really a dynamic and evolving decision-making process (Veeder 1975).

This summary of four models indicates that there is certainly a progression, with the most recent being the most structurally sophisticated but probably the hardest to use empirically. There are, however, problems in comparing one with another due to differences in intent and idiosyncrasies in definitions. Nevertheless, each is in a sense cumulative, building upon the earlier models, and a greater precision seems to be developing. Gross's model is particularly interesting, attempting as it does to predict and explain utilization behaviour and, if operationalized, this could be a most promising advance. As Stimson (1981) reports, attempts continue to be made to bring this model into use. For the geographer and health service planner, however, all four of these models still represent more of an attempt to specify the nature of variables influencing utilization rather than a clarification of the effects of each. Perhaps the greatest stumbling block to effective application is that, as in the majority of models, the normative assumptions upon which they are based will limit their explanatory power in what is essentially a dynamic behavioural arena.

A useful basis for examining health systems and their utilization is provided by Aday and Andersen (1974). Although proposed as a framework for the study of 'access' to health care, it really involves a wider proposition: that national health policy should be taken into account when viewing the delivery of health care by a given system for a given population at risk. These provide the INPUTS for the model (Figure 6.1). The OUTPUTS are the utilization of health services, which is one measure of the system's adequacy for the population it serves. This may be regarded as an 'outcome' of the system, and with it comes a level of 'satisfaction'. This could be measured in terms of percentage of a study population who were satisfied with specific aspects of the service used such as quality, convenience, cost, co-ordination, and courtesy.

The model provides a useful framework for researching the mechanics and outputs of any health system. It is suggestive of the political economy approach discussed in Chapter 2 since national health policy is viewed as having overall influence on the inputs and outputs. Moreover, the framework provides a research design for the collection of empirical data by survey in any given health care system. It is also valuable as it recognizes that access involves more than the mere existence or availability of resources at a given time; indeed, 'geographic accessibility ... refers to the friction of space that is a function of the time and physical distance that must be traversed to get care' (Aday and Andersen 1974, pp. 209-210). this will form an important recognition later in this chapter when the effects of distance decay are discussed for specific services.

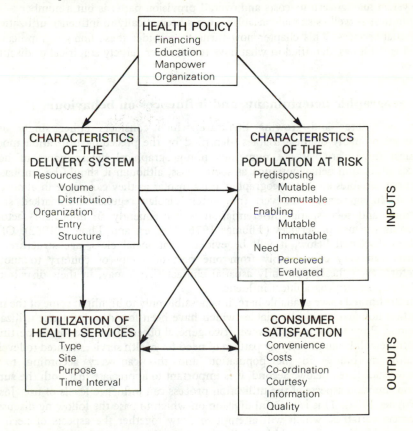

Figure 6.1   A framework for the study of access to health services

Source: Aday and Andersen (1974)

Aday and Andersen also suggest that 'access' may be gauged by the utilization of services by designated populations for whom they are intended. However, they further suggest that access may be even more appropriately considered in the context of whether those persons actually in *need* of medical care receive it. The fact that many go untreated who would ideally be treated by a physician can be considered something of a 'medical iceberg', in which the greater the proportion above the water, the greater the access to care of the group represented by the iceberg (Beck 1973).

Taken as a group, these models provide a valuable contextual fabric for the geographically oriented models introduced in the previous chapter. They are of relevance for human services planning because, in spite of individual failings and a general lack of dynamic-temporal considerations, they indicate to researchers the range of potential variables which need to be considered when analysing the utilization of any given health service in any given health care system (Veeder 1975). The same basic variables would require collection and analysis when looking at the pattern of utilization in any of the nations discussed in Chapter 2. The aggregate characteristics of a system may govern its costs and overall provision patterns but a number of non-geographical as well as spatial variables seem systematically to influence utilization of individual services. This chapter now turns to consider these non-geographical and spatial variables as identified in what have to date been largely empirical or discursive studies.

Non-geographic determinants and influences on behaviour

There is a large and growing literature which considers either singly or in combinations many of the variables identified by the preceding utilization models. Amongst these variables are numerous non-geographical determinants of health services utilization behaviour such as social class, although it should be emphasized that these variables are 'non-geographic' only insofar as they can occur in almost any location. In aggregate, however, they often display regular and marked spatial variations, and socio-spatial differentiation is consequently an important factor in examining the use of services (Thomas 1976; Herbert and Thomas 1982). Of the variables discussed below, it will be evident that social class, age structure, and ethnicity can vary considerably from one part of a city or country to another. Therefore, even these apparently aspatial characteristics may, in their distribution, acquire important geographical influence.

In the limited space available here, it is possible only to highlight some of the more important non-geographical variables which have been found to influence utilization behaviour. Prior to doing so, however, two general observations are highly pertinent. First, there will be an underlying pattern of need for health services related to levels of ill-health (morbidity) in the population and this can vary according to the characteristics listed below. Second, it is important to appreciate that both the supply and consumption aspects of the utilization process can influence levels of use (Joseph and Poyner 1982). This is a useful division on which to base the following discussion of discrete variables, which will attempt to draw together the aspects of need and utilization. However, it should be realized that the discussion of discrete variables as influences on utilization is, in some ways, unrealistic and an oversimplification. Rarely will variables operate singly, more usually they will act in combination to influence utilization. Indeed, it is the problem of distinguishing individual effects that has for so long hampered clarity of explanation in this research topic. The selection of variables for discussion is therefore somewhat arbitrary but reflects their apparent importance at present.

Social class

Perhaps one of the most pervasive influences on utilization behaviour is social class of the consumer and, sometimes, of the provider also. It is difficult to distinguish clearly its effects, though, because definitions of 'class' vary internationally and from study to study. There are also inherent problems in establishing its effects because, as a variable, it is intricately connected with occupation, income, status and education. However, studies within Britain have often used the Registrar General's classification (Office of Population Censuses and Surveys 1970), whilst those in North America have used indices such as socio-economic status measured by combining ranks on education and occupation (McBroom 1970).

In many countries, studies have shown that the lower social class groups (however defined) tend to have a greater *need* for health care, associated with higher levels of morbidity and larger families (Hart 1971; Salkever 1975; Pyle 1979; Department of Health and Social Security 1980; Whitelegg 1982). Figure 6.2 shows that in the United Kingdom, for instance, there seem to be generally increasing rates of ill-health, both chronic (long-standing) and acute, as lower social class groups are reached (Foster 1976; Department of Health and Social Security 1980).

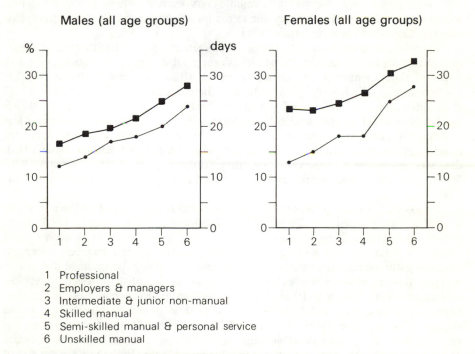

1 Professional
2 Employers & managers
3 Intermediate & junior non-manual
4 Skilled manual
5 Semi-skilled manual & personal service
6 Unskilled manual

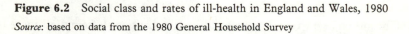

■——■ days restricted activity per person per year
●——● chronic illness:- percentage of respondents reporting limiting long-standing illness

Figure 6.2 Social class and rates of ill-health in England and Wales, 1980

Source: based on data from the 1980 General Household Survey

However, do utilization *rates* vary significantly between and amongst social classes, with the 'more needy' frequently losing out? This has developed into something of a contentious debate, particularly in Britain. Depending on which data are used and which medical care services are considered, lower social class groups can be shown to use health services more (Rein 1969; Foster 1976) or less than other social classes (Cartwright 1967; Titmuss 1968; Alderson 1970; Blaxter 1976; Cartwright and O'Brien 1976). Moreover, there is evidence that changes in utilization rates occur over time, which means that it is difficult to extrapolate findings far beyond the time of the research. Comparison of the results of two important surveys of general practice in Britain, conducted in 1964 and 1977, indicates that considerable changes have taken place in the utilization patterns of certain social groups (Cartwright 1967; Cartwright and Anderson 1981). Whilst overall GP consultation rates had risen little (from 3.4 to 3.6 per person annually between 1964 and 1977), changes had occurred in the rates at which working-class men were seeing their doctors; they had increased their contacts to rates similar to those of middle-class men during the thirteen-year period and they seemed to be receiving increasing amounts of preventive and screening services associated, perhaps, with enhanced occupational health services. Another increased utilization pattern was for younger, middle-class women, whose consultation rate increased from 3.2 to 4.2, probably due to the increased use of GPs for contraceptive services (Cartwright and Anderson 1981).

Even setting aside class differences in utilization rates of health care, it seems certain that lower-class persons do not get as much out of their consultations, either in terms of time spent with the doctor or advice (Buchanan and Richardson 1973; Cartwright and O'Brien 1976). In particular, there is evidence of considerable social class differentials in the use of preventive health services (Sansom et al. 1972; Waddington 1977). In most studies, lower social class persons are found to receive less health information and promotive care; examples include ante- and post-natal care (Cartwright 1970; McKinlay 1970; Brotherston 1976), family planning (Bone 1973), dentistry, particularly conservative dentistry (Bulman et al. 1968; Alderson 1970; Waddington 1977), and health education received through preventive medicine (Cust 1979).

Socio-economic status has been identified as influencing use of services in many countries other than Britain (Department of Health and Social Security 1980). Koos (1954), in a classic New York study, found that lower-status respondents were frequently deterred from seeking medical care because of cost and fear, although evidence garnered elsewhere suggests that it should not always be assumed that lower-income (lower socio-economic status) persons underutilize physician services (McBroom 1970; Galvin and Fan 1975; Cockerham 1978). Sometimes, low-income families even fare quite well in terms of access to services and may be inclined to use them if they are 'free' (Monteiro 1973; Salkever 1975). As yet, however, there is relatively little work which unambiguously 'explains' utilization in terms of social class differences, the explanatory potential of much research being limited by the paucity of comprehensive data on use of health services by individuals of different social class.

Income

The effects of income on its own are, of course, difficult to distinguish from those more broadly attributable to socio-economic status. It was previously suggested that in the United States lower-class persons underutilize health services because of the financial costs of services and because of a subculture of poverty that has failed to emphasize the importance of good health (Cockerham 1978). McKinlay (1972) points out that there are a number of examples of economic barriers to the receipt of care in fee-for-service systems. In particular, high costs of medical care and low incomes appear to force some groups to change their venues of health care. In a study in Oklahoma City, for example, the elderly were identified as attending hospital emergency rooms and outpatient departments for primary care due to the cost of private physicians (Bohland and French 1982). In countries such as Britain, Russia, and China, which have 'national' health care systems, economic barriers to utilization should in principle be limited only to the costs of transport to service. However, in Britain, a growing private sector in medicine, which many cannot afford, and perhaps increasing waiting lists for public health care, may be creating some economic barriers even within the NHS, as discussed in Chapter 2 (Politics of Health Group 1982).

In addition, it is quite possible that income only indirectly affects utilization and underutilization (McKinlay 1972). Undoubtedly, some lower-income regions are poorly provided with medical services (see Chapter 4) and this, in itself, may cause lower levels of utilization because of inferior availability. More important, perhaps, in a fee-for-service system, will be insurance status. In Australia, prior to the Medibank experiment of 1975, between 10 and 15 percent of the population were not insured, most of them were drawn from low-income groups and were discouraged from taking up insurance by frequent rises in premiums (Goldstein 1982). As a result, utilization of many aspects of the service for which fees were charged directly was curtailed, an experience noted in many other similar systems.

Age and sex

These are two manifestly important variables influencing health services utilization. It may be expected that, as they become older, and especially after retirement, people will need health services more for chronic and perhaps for acute conditions. However, among younger age groups, utilization by females can be increased for reasons associated with childbirth and contraception, so that their use of services will almost always tend to be higher than that of males. Numerous studies have found that women have higher morbidity than men (Anderson and Andersen 1972; Kohn and White 1976; Verbrugge 1979; Cleary et al. 1982) although this obviously varies very much according to cause (Roberts 1976; Cust 1979), men, for example, not experiencing the physical trauma which can be associated with childbirth!

The underlying reasons for differences in morbidity are varied and often unclear although they probably include genetic, environmental, and occupational factors. Males are generally more prone to cancers except those of reproductive organs (Ashley 1969a, 1969b) although they are usually less prone to psychiatric disorders. This has been supported in Britain, where 11 percent of males but 17 percent of females are estimated to experience a psychiatric disorder at some time during their lives, and in the United States, where women in 1966–1968 comprised at least 60 percent of the adult population in psychiatric facilities and two-thirds of diagnosed schizophrenics (Roberts 1976). Davey and Giles (1979), on the other hand, found a reversal of the picture in Tasmania where there were generally more men than women receiving psychiatric treatment in hospital, perhaps because of more home treatment of depression or neuroses. However, females have consistently been found to have higher rates of depression than males although the reasons for this are not clear (Rosenfield 1980).

Nevertheless, in spite of (possibly) greater morbidity, women outlive men on average, perhaps mainly because of higher incidence of accidents and occupational mortality amongst men. This difference has generally been increasing this century in developed countries and it can have a number of social consequences in terms of the care for elderly widows. Table 6.1 shows this feature for the United States, England and Wales, and New Zealand up to 1965. It has, by and large, continued subsequently (Department of Health and Social Security 1980).

Table 6.1   *Expectation of life at birth: differences between males and females*

		Years		
		M	**F**	**(F–M)**
United States	1910	48.46	52.01	3.55
	1930	57.31	60.70	3.39
	1950	65.30	70.92	5.62
	1965	66.88	73.87	6.99
England and Wales	1911	49.35	53.38	4.03
	1930	59.02	63.27	4.25
	1950	66.57	71.32	4.75
	1965	68.53	74.79	6.26
New Zealand	1911	60.27	63.12	2.85
	1930	64.36	67.26	2.90
	1950	68.39	72.22	3.83
	1965	68.78	74.65	5.87

Source: after Roberts (1976)

The effects of increasing age and of sex differences on utilization are therefore difficult to generalize. Perhaps women use services more than men simply because they generally have more knowledge about health matters (Feldman 1966). Indeed, measures of perceived health, days off work, and health conditions are all problematic since they can reflect differential knowledge and perceptions of health or illness. More recent studies seem to confirm that age-sex differences need to be examined in relation to perceived health and help-seeking tendencies, and also in relation to different types of physical use: discretionary, physician-initiated, and preventive. For example it seems that, in the developed world at any rate, women tend to receive more 'total services', more extensive diagnostic services and more follow-up consultations (Verbrugge and Steiner 1981). Overall, they seem to have more frequent contacts with physicians (Cockerham 1978), although it may be that this is due to visits connected with childbirth or contraception. However, in a large study in Wisconsin, involving user interviews and use of medical records, women who had given birth in the year before the interview were removed from the sample, but this only slightly reduced aggregate female utilization rates to 2.88 visits to a physician per year compared to 1.94 for men. This study suggested that sex differences in utilization can be attributed to *real* differences in health, in spite of females' greater longevity (Cleary et al. 1982). On the other hand, the sex variable *may* be significant only to the extent that women *report* more illness than do men and 'the masculine role probably dictates a "business as usual" attitude even though a man may feel sick' (Cockerham 1978, p. 77). It is therefore evident that even apparently straightforward differences in utilization may be more apparent than real.

It is fair to say that persons over the age of 60 or 65 will tend to be in poorer health and to be hospitalized more frequently than other age groups (Anderson and Andersen 1972) and that physical competence generally decreases with age (Birdsall 1979). However, this will vary very much among individuals. Phillips (1981a), for example, in a study of general practice use in South Wales, Britain, found an apparent dichotomy amongst people over 65 years of age between 'regular' or frequent attenders and non-attenders (presumably the old but healthy). In a number of North American studies, it has been confirmed that elderly persons are more likely to visit a physician than younger age groups except the very young (Donabedian 1973; Monteiro 1973; Galvin and Fan 1975; Joseph and Poyner 1982). However, they may not necessarily visit doctors as often as they should do in the light of their symptoms and not as often as would be desirable (Taylor et al. 1975). Table 6.2 illustrates steadily increasing consultation rates in Britain as age progresses, but with differences between men and

Table 6.2 *Average numbers of consultations with a GP per person per year in England and Wales*

Age	Men		Women	
	1971	1977	1971	1977
15–44	2.4	2.4	4.5	4.3
45–64	3.4	3.6	4.2	3.9
65–74	5.0	3.8	5.3	4.2
75+	6.8	6.1	7.4	5.1

Source: Cartwright and Anderson (1981), based on General Household Survey data

women's utilization especially for those over 65 years. The generally falling rates of contact between 1971 and 1977 are, it is suggested, due to fewer home visits by doctors and an increasing number of day hospitals, which reduced the need for contact with the GP (Cartwright and Anderson 1981).

These statistics tell only part of the story, however. At the level of the individual health provider, an ageing population translates into increasing pressure on available resources. In a general practice in Liverpool, England, heavy work-load for doctors was found to be caused by elderly persons, especially women, in part because of their larger numbers in the age group. Patients over 64 had a new call rate of 2.3 times the average whilst their repeat visit rate was 3.9 times the average (Hopkins et al. 1968). Additionally, in many studies of both hospital and primary care, the elderly have been seen to produce higher consultation rates (Galvin and Fan 1975; Walmsley and McPhail 1976; Haynes and Bentham 1982). In view of the increasing proportion of elderly persons in most developed and some developing countries, their apparently greater pressure on health services is very important in planning for health care (Chapter 8).

Ethnicity

The cultural or ethnic background of individuals has been suggested as a cause of differential utilization behaviour in health services. In the United States, for example, it is suggested that preventive medicine is very much a white, middle-class preserve which racial/ethnic minorities will use less. Frequently, these minorities may make use of paramedical healers, pharmacists, friends, or a variety of 'fringe' healers, turning to a professional physician only quite late in their illnesses, having been through a 'lay referral system' (Freidson 1970; Hines 1972). This behaviour may be closely connected with traditional health beliefs in which disorders are seen not to require the intervention of 'scientific' medicine but to need instead the restoration of 'balance' in the body. In Chapter 2, it was seen that many indigenous health systems exist in Africa, Asia, and Latin America, and the health behaviour of minorities in Western societies, for instance, Mexican-Americans or New Commonwealth immigrants in Britain, may also be founded on such traditional beliefs imported in the 'cultural baggage' of immigrants.

Within Western countries, ethnicity can underlie access to inferior health facilities due to residence in less desirable locations (De Vise 1973; Shannon and Dever 1974; Thomas 1976) or because of linguistic or cultural impediments. Therefore, directly or indirectly, it can be a cause of barriers to utilization. Underutilization may be related to poverty (the inability to afford treatment), a lack of faith in the system to provide for ethnic minorities, or poor communications between doctor and patient. This matter is therefore closely bound to income and social status variables, as ethnic minorities in any country are often amongst the poorer groups. Within the poorer ethnic groups, the elderly or those with large families may be least able to use health care effectively, compounding other variables discussed earlier. In addition, it is possible that some health care facilities will be denied to ethnic minorities purely because of their race, so explicit 'racial barriers' may exist to utilization.

The preceding discussion of some apparently non-geographic determinants of utilization has, because of space constraints, omitted or glossed over potentially important and often unquantifiable factors. These include lifestyle (for which social class is often used as a surrogate measure), diet, health status or disability, occupation, and direct measures of income. In addition, the social-psychological approach to explaining utilization, in which motivation, perception, and learning are deemed to be key elements in determining utilization behaviour, has largely been neglected (McKinlay 1972). Perhaps what should be sought is clearer knowledge of social-psychological factors and the ways in which they relate to other variables in their influence on the utilization of medical services (Mechanic 1968; Cockerham 1978). Unfortunately, such developments are still some way off.

Distance decay effects in utilization

In the previous section, specific variables were discussed which have been identified as influencing the use of various types of health services. Sometimes they appear to act independently; at other times, the individual effects of variations in age, sex, income, and social class are difficult to distinguish. However, geographers are convinced that spatial factors have very important roles to play in the use of health services. These can be aggregate, reflecting differential regional availability of services, or they can be more personal, reflecting the local accessibility of health care facilities to individuals whose personal mobility varies. Sometimes, these spatial factors may be the key variables determining whether an individual will or will not use a health service. Unfortunately, when measuring utilization by means of interviews or researching health care provider records, 'attendance' rates are usually discovered but it is notoriously difficult to establish rates and reasons for *non*-attendance.

When examples of distance decay effects are discussed below, it must be acknowledged that *patterns of revealed behaviour* are being interpreted. It does appear that most studies in the past have confirmed the existence of distance decay in the use of physicians and hospitals, as their use rates generally vary inversely with distance (Giggs 1983a). However, these have by no means been found to have been uniform rates of decay, as numerous factors have distorted patterns, including size of facility, catchment area, range of services offered, and transport availability and costs, not to mention the variables outlined in the previous section. Nevertheless, geographers have felt confident in using the distance decay concept in modelling intraregional accessibility to health care (see Chapter 5).

It has become increasingly obvious that behavioural explanations of consumer health behaviour must be sought. This understanding has developed considerably since the mid-1970s, helped in part by advances made in the modelling of consumer behaviour in other, non-medical services (Thomas 1974, 1976; Foxall 1977). What has emerged most clearly from the subsequent medical work in this field is that distance decay does not appear to operate uniformly for all types of services, nor in all countries. A further complication is that the effect of distance can apparently be enhanced or reduced by the nature of symptoms or illness which patients are suffering (Girt 1973). As a result, most empirical studies to date have been surveys of specific aspects of consumer behaviour within particular sectors of the health care system in given countries (Giggs 1983a). This is a fair comment and it is also based strongly on the commonsense recognition that the stimulus to attend medical services will vary greatly, for example, between a non-urgent visit to the dentist or doctor and an emergency visit to accident and emergency facilities. Therefore, the severity, possible outcome, and perceived importance of the 'utilization episode' (in jargon terms) will probably all affect the influence of distance from source of care on use rates. Very few surveys have attempted to examine consumer health care behaviour across the whole spectrum of health services and those that have done (such as Krupinski and Stoller 1971) have, as Giggs (1983a) points out, generally lacked strong geographical foci. There are now appearing important empirical studies which take into account behaviour with regard to a range of health services although unfortunately these are still relatively few (Cleland et al. 1977a; Haynes and Bentham 1982).

The remainder of this section will focus on specific sectors of health care systems to provide empirical evidence of distance decay effects in utilization in a variety of national settings. It is accepted that this approach will be somewhat fragmentary but the value to the student and researcher of empirical studies to which to relate new findings is probably considerable. At this stage, discussion is delayed for one part of the health care system in which much pioneering work has been done on the effects of distance; that is, mental ill-health care. This forms the subject matter of the following chapter.

The studies considered will include services drawn from the secondary as well as the primary tier of the health care hierarchy. It is unwise, however, to think of these levels of health care as being mutually exclusive (Chapter 3). For example, it is obvious that emergency or accident facilities are used by some as primary care facilities (Cartwright and Anderson 1981) and they may become the main source of care for many persons, particularly the aged and minority and indigent groups (De Vise 1973; Pyle 1979; Roghmann and Zastowny 1979; Giggs 1983a). At the other end of the medical care spectrum, visits to some facilities such as the dentist may be regular but infrequent events or may take on emergency characteristics. The nature of service is therefore again stressed as a key variable in analysis.

Emergency departments

Visiting rates for hospital emergency departments appear to have been increasing steadily in recent years (Conway 1976; Davidsson 1978; Steinmetz and Hoey 1978; Magnusson 1980). This may be attributable to the fact that these are often the only free or subvented sections of many health systems. In other systems, such as in the USA, most hospitals charge even for emergency treatment.

The increasing use of hospital emergency departments seems to be explained by a number of factors, such as reduced availability of private and family physicians out-of-hours, especially during the evening and at weekends (Magnusson 1980; Phillips 1981a). 'Unavailability' may be perceived rather than actual, but its effects can be very real nevertheless. Attitudes have also changed, many patients being aware that injuries due to accidents often need diagnostic and treatment facilities only available in hospital and that they will be referred there even if they see a primary physician first. Others realize that in many cases accident and emergency room visits may provide one way of circumventing queues for specialist attention. Finally, for medico-legal reasons, most emergency departments find it difficult to refuse to see patients and tend to investigate quite thoroughly apparently everyday complaints in case of missing a serious disease or a complication of an accident (Magnusson 1980). This is in spite of the fact that, in Britain, for example, casualty departments are entitled to refuse to see non-emergency cases; advice rarely adhered to. As a result, hospital emergency services tend to provide true emergency and trauma treatment *plus* back-up services for off-duty physicians and a substitute for no care (Torrens and Yedvab 1970).

Distance decay effects have been noted in studies of use of emergency room facilities in a number of countries (Ingram et al. 1978; Roghmann and Zastowny 1979; Magnusson 1980). Some use 'historical' sampling of ER (emergency room) visits from hospital records over a number of years; others concentrate on shorter time-scales but, sadly, true behavioural data are generally lacking on this topic. Nevertheless, a particularly useful study is that by Magnusson (1980) of the visits to Huddinge Hospital Emergency Department in Stockholm County, Sweden, between January 1976 and March 1977. This new University hospital had a defined catchment population with access only to this one emergency department and, apart from utilization data from the hospital, extra socio-demographic data were obtained through the Stockholm County Medical Information system. The objectives of the study were to investigate a 30 percent increase in Stockholm in ER visits between 1973 and 1977, and to research the relationship between travelling distance and visiting rates to the hospital emergency department, plus subsequent use of outpatient and inpatient care. Magnusson's hypotheses were that visiting rates would decrease with distance from the hospital and that there would be a positive correlation between visiting rates to the emergency department and the use of outpatient clinics and inpatient care.

The data set was a 6.4 percent sample of visits during the 15-month period up to March 1977, a sample of some 9,632 visits. Socio-demographic data and home addresses of patients were available and it was possible to assign each to a subarea within the hospital's catchment, from which distance and travel time to the hospital were calculated.

The results indicated a significant negative correlation between visiting rates and travelling distance and there was also a significant positive correlation between the proportion of new immigrants to the County and visiting rates (Table 6.3). The distance decay relationship for the 20 catchment subareas is displayed in Figure 6.3. The distance decay model accounted for 68 percent of original variation in visiting rates but when the proportion of immigrants was included in the analysis (that is, proportion non-Swedes), an expanded model explained 81 percent of variation in visiting rates amongst the 20 subareas.

This is one of the more remarkable examples of a distance decay effect although, as the author recognizes, it is not perfect, since personal mobility, time of day or week, and urgency of medical problems are not taken into account. The second hypothesis also seemed to be upheld; there did appear to be a positive correlation between

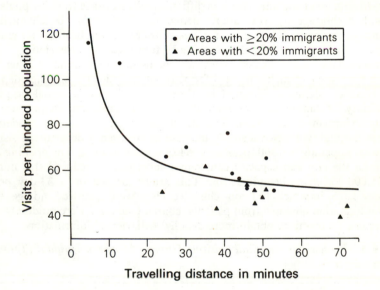

Emergency department visiting rates for each subarea plotted against travelling distance to the emergency department with the function:

$$Y = A + \frac{B}{X} \text{ fitted to the data}$$

Subareas with high and low immigrant proportion are indicated

Figure 6.3 Proximity and the use of hospital emergency departments in Stockholm

Source: after Magnusson (1980)

Table 6.3 *Correlations between emergency department visiting rates and other variables, Stockholm 1976–1977*

	Mean age	Immigrant proportion	Travelling distance	Visiting rate
Mean age				
Immigrant proportion	−0.5116*			
Travelling distance	0.0732	−0.3928		
Visiting rate	−0.2206	0.5754**	−0.7889***	

* Significant at the 0.05 level   ** Significant at the 0.01 level   *** Significant at the 0.001 level

Source: after Magnusson (1980)

emergency visiting rates and utilization of outpatient and inpatient care. It is difficult to explain this relationship precisely, but it is perhaps influenced by enhanced knowledge of illness, increased referral, or by the unavailability of GPs. Whatever the reasons, Magnusson concludes that it is difficult to divert usage from hospitals and, in areas with a shortage of GPs, unmet demands will probably be channelled into emergency departments. This, of course, has considerable implications for emergency department work-loads and for the planning of future facility provision.

Distance decay in the use of emergency rooms has been found in other studies, although not always in such a marked form. In an analysis of emergency department use at Humber Memorial Hospital in Metropolitan Toronto, actual or perceived unavailability of family doctors seemed to be an important factor encouraging attendance (Ingram et al. 1978). The study, based on a relatively modern acute treatment hospital, witnessed a steady increase in emergency department visits since the hospital's opening in 1960, peaking at 54,000 visits in 1972. The opening of other hospitals in the city and capacity constraints subsequently reduced this number to some 45,000 visits in the mid-1970s. This study focused on a 33-day period in May–June 1975, in which 400 non-scheduled visits were sampled, and age, sex, and address information obtained from patients' charts. One-third were randomly selected for telephone follow-up to obtain more detailed behavioural information.

Table 6.4 *Reasons for specifically visiting Humber Memorial Hospital (Toronto) emergency department*

Reason	Frequency	Percentage
Closest or no time	23	25.0
Recommended	11	12.0
Influence of doctor(s)	27	29.3
Previous association	25	27.2
Just passing by	4	4.3
Prefer HMH to others	2	2.2
TOTAL	92	100.0

Source: after Ingram et al. (1978)

Table 6.4 indicates that a number of factors seemed to influence the patients who specifically chose Humber Memorial Hospital, and 25 percent viewed the perceived temporal or spatial accessibility of the hospital as important. The catchment area of this urban hospital was comparatively small because 90 percent of the patients were drawn from a radius of 3.6 miles. The average straight-line distance patients travelled was 2.4 miles although such a measure does not take into account either travel time or actual distance travelled. As in the Swedish study, the number of visits by subarea had a negative exponential distribution, suggesting that utilization is strongly related to distance from hospital (Figure 6.4). This is probably especially true in urban settings, such as Toronto, in which there will be a number of 'competing' emergency departments available. However, Ingram et al. (1978, p.60) write that 'it may be postulated that the visibility of the hospital, or that patient awareness, declines with increasing distance'. This seems very likely in the intrametropolitan setting.

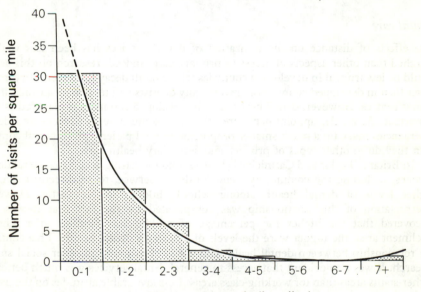

Figure 6.4   Distance decay effects in the use of emergency department services, Humber Memorial Hospital, Toronto

Source: after Ingram et al. (1978)

Distance decay effects do not always appear to act uniformly for all patients. A study of some 8,600 ER visits to six hospitals in Monroe County, New York, sampled from a six-year period up to 1974, indicated that small, isolated, rural hospitals, and those serving patients of low socio-economic standing, tended to display the steepest slopes of distance decay in utilization (Roghmann and Zastowny 1979). In hospitals known to serve patients of higher socio-economic standing, the effects of distance on utilization were notably less severe. As the authors note, for higher-status patients, travel distance did not seem to be a barrier, because such patients would apparently travel a long distance rather than give up their preferred hospital environment.

Although most evidence is clear on distance decay effects in ER visits, they have not been noted in as distinct a form in research in rural western New York and north-western Pennsylvania (Basu 1982). As noted above, several studies have suggested that low-income families are reluctant to travel long distances for medical care (Kane 1969; Roghmann and Zastowny 1979). However, Basu observed a substantial number of low-income families were travelling long distances to ER services. Perhaps rural patients are willing to travel greater distances for medical care, in search of suitable services regardless of the presence of physicians or hospitals in their communities. The larger, better-located hospitals, presumably employing better physicians, seemed to attract emergency room patients from relatively distant areas. Patients were, perhaps, avoiding smaller hospitals which were busy, with long waiting periods. Instead, some were apparently undertaking long journeys to reach better care which adds the variable of quality and size to the study of this particular distance decay phenomenon!

Dental care

The effects of distance on the utilization of dental services has been less well re-searched than other aspects of access to primary care. Indeed, research on this aspect would be less fruitful in developing countries where tooth decay rates seem very much lower than in developed countries and where any dentists will tend to locate only in the largest centres. However, dental health care in developed countries is regarded as very important. As a rule, appointments are scheduled some time in advance and dental emergencies make up a much smaller proportion of total patient–professional contacts than they do in other types of primary and secondary health care.

In Britain, Taylor and Carmichael (1980) found increasing access to general dental services, including the community (schools) dental service, to be directly related to higher levels of dental health among school children in Newcastle. However, interpretation of this relationship was complicated by the fact that it was also discovered that the higher the percentage of middle-class families in a school's catchment area, the higher were the levels of children's dental health. This confirms the relationship betweeen dental health and social class found in other social survey research. It was also evident that access to general dental services was much better for higher-status areas than for working-class areas. Less favourable attitudes on the part of lower-status persons to dental care could perhaps explain part of the difference (Dickson 1968) and this could be compounded by the fact that there are fewer dentists in lower-class areas (maybe as a result of dentists finding less work or fewer patients in such locations, of course!). Spatial analysis will tend to concentrate on access to services and its effects on dental health, and thus illustrates a locational process influencing dental health which social surveys can fail to uncover (Bradley et al. 1978; Taylor and Carmichael 1980).

Whilst not directly measuring the effects of distance, the influence of differential availability on the uptake of dental treatment by school children has been noted. O'Mullane and Robinson (1977) studied two towns in north-west England which had widely different population:dentist ratios, town **U** having approximately twice as many patients per dentist as town **F**. Clinical examinations of the school children were conducted and two indices of dental health calculated: a filling:extraction ratio (F/E) and restorative index (RI). For both indices, a higher value indicates better dental health. Table 6.5 shows the differences between towns **U** and **F** and the differences

Filling:Extraction ratios (F/E) and Restorative Index (RI) of 14-year-old
ʰy availability of dentists and social class, North-West England

	Town U		Town F		Differences	
	1:4,631		1:2,505			
	F/E	RI	F/E	RI	F/E	RI
ᴊcial class I & II	5.2	85.3	7.7	90.5	2.5	5.2
₂ 2. Social class III	2.5	63.3	6.5	84.9	4.0	21.6
ᴊroup 3. Social class IV & V	1.4	63.4	6.5	81.7	5.1	18.3
TOTAL SAMPLE	2.5	71.3	6.9	85.8	4.4	14.5
Differences (Group 1 – Group 3)	3.8	21.9	1.2	8.8		

Source: after O'Mullane and Robinson (1977)

between social classes in each. The availability of dentists to the different social groups appears to be fundamental. Where the population:dentist ratio is more favourable (Town **F**), not only does Social Group 3 take up more treatment but dental health improves. Since the availability of dentists seems to be of greater significance than sociological attitudes to dental care in explaining differential uptake of treatment, O'Mullane and Robinson (1977) suggest that a means of effecting improved dental services uptake is to increase the output of dentists willing to work in underprovided areas. This evidence is more positive than in the case of medical care and health levels. As Taylor and Carmichael (1980) and Carmichael (1983) also conclude, redistribution of resources will probably have more immediate effects on dental health than a policy of changing attitudes through dental health education. This recognition may be crucial to investment and training policies which at present place considerable emphasis on health education to prevent dental decay. Perhaps efforts to reduce the 'friction of distance' by improving accessibility would be more effective, if more costly.

Shannon et al. (1973) have emphasized that both travel time and travel distance need to be taken into account when investigating the 'friction of space' for dental and other health care service utilization. In spite of shorter travel distance for medical care in a Cleveland study, inner-city residents had to travel for longer times than their suburban counterparts. Therefore, whilst in distributional and straight-line accessibility terms, inner-city residents appear favoured, in reality, they appeared to be disadvantaged in seeking medical care, which must, it is concluded, influence rates of use of the services.

Joseph and Poyner (1981, 1982) also illustrate distance decay effects for the use of both medical and dental services in rural Ontario. However, they question why location of residence influences the pattern of medical service utilization. Perhaps the differences between the profiles of users of local and non-local facilities result from the spatial distribution of household income, age, social class, and other variables, rather than distance per se. This can only be answered by a more explicitly behavioural level of analysis. We shall return to this observation at a later point.

General practitioners or primary physicians

The first line of contact with medical services is generally the family doctor (primary physician or general practitioner). Countries vary immensely in the degree of continuous contact with the same doctor that is usual and this in itself may influence utilization. Numerous examples exist of the influence of distance on the use of GPs although few have explicitly examined distance decay functions, owing, mainly, to the problem of gaining access to practice records. The size of GP practice also seems crucial. In Adelaide, South Australia, for example, Stimson (1981) found the 'trade area' for most solo GPs to be extremely restricted, with 80 percent of customers coming from within a 2 km radius. For large group practices of three, four, five or more doctors, catchment areas extending beyond 5 km were observed. Very different effects of distance on utilization could be expected in each type of practice, making generalization difficult.

However, some trends can be detected. In spite of a selective retention of patients moving more than 2 miles (3.2 km) from surgery, in Liverpool, a distance decay effect was noted for patients living more than 1 mile (1.6 km) away, both for surgery and home consultations (Hopkins et al. 1968). In particular, the physicians found themselves less likely to make repeat visits to homes of patients living more than 1 mile from the surgery. Conversely, patients living within 1/4 mile of the surgery had a higher surgery visiting rate than the more distant residents. In a study of a general practice in London, the consultation rate of patients was inversely related to distance from the surgery (Morrell et al. 1970). In a secondary analysis of the data from this study, Parkin (1979) detected more subtle effects of distance on different groups of patients. Consultation rates on aggregate fell from 5.0 consultations per head of population per year for patients living at less than 5/8 mile (1 km) from surgery to 3.6 consultations for those living at a greater distance. Home consultations also fell, as reported by Hopkins et al. (1968), at a similar rate.

Parkin (1979) distinguished between overall consultation rates (every occasion on which a patient received professional advice or treatment from his doctor) and attendance rates. This showed that all groups of patients but one (men aged 15–64) showed significantly lower consultation rates as distance increased. For actual attendance rates, distance seemed to affect most those whose use rates are relatively high: females, the elderly, and those in Social Classes III, IV, and V. The elderly were particularly influenced and show most clearly the negative effects of distance on utilization (Table 6.6). This trend is worrying, of course, as it may mean that people in real need of primary health care and medical monitoring are avoiding seeing the doctor due to increasing distance. It is important that health planners take note of such findings because, as the size of practices increases (Chapter 4), areas of deficiency of service provision may be caused by the greater catchment areas and distances to surgery, especially for needy groups (Phillips 1981a).

Parkin (1979, p. 97) was able to suggest that 'the distances from patient homes did deter the population both from consulting at all with the doctor, and also from attending regularly'. Again in Britain, another study of some 400 respondents found relatively little evidence of distance decay in *place* of attendance, that is, quite large numbers of respondents were attending surgeries other than those near to their homes as Central Place Theory would suggest (Phillips 1979a). However, there was some evidence of distance decay in utilization *rates* since respondents living within 2 miles of

Table 6.6   *Annual consultation rates by distance to surgery, and age, sex and social class groups*

	Overall	Zone 1 (<0.25 miles)	Zone 2 (0.25 to 0.625 miles)	Zone 3 (>0.625 miles)
Population groups				
Whole population	4.82	5.07	5.18	3.53
Males	4.17	4.17	4.80	3.37
Females	5.40	5.89	5.53	3.67
Social Class I and II	4.14	4.29	4.98	2.45
Social Class III, IV, V	4.88	5.14	5.21	3.63
Age groups (years)				
0–14 male	4.18	4.16	5.17	2.88
0–14 female	4.55	4.90	4.59	3.35
15–64 male	3.73	3.57	4.33	3.46
15–64 female	5.22	5.52	5.52	3.91
Over 65 male	7.08	8.20	6.75	3.86
Over 65 female	7.49	8.97	7.12	3.02

Note: Distances groups are from patients' homes to surgery. All differences of Zone 1 and Zone 2 from Zone 3 are significant at the 0.001 level except 15–64 male age group.

Source: after Parkin (1979)

surgery did attend rather more frequently than those living further away and frequent attendance was apparently curtailed by greater distance (Phillips 1981a). This implies that only the more urgent types of cases may attend surgery if living further away. The converse of this seemed to be that more distant residents received more home visits in some survey sites. However, since this was not a well-developed trend, it was difficult to suggest that more distant patients placed greater demands on the family doctor, supporting the conclusions of Hopkins et al. (1968). Nevertheless, this study emphasizes the need to distinguish between *attendance* at a source of care and *frequency* of utilization of that source.

There are relatively fewer examples of geographical research into the use of primary medical services in countries outside Britain. Notable exceptions are by Cleland et al. (1977a) in Adelaide and by Joseph and Poyner (1981, 1982) in rural Ontario. In the latter study, distance decay was evident in the use of a specific medical centre. However, for service use (as opposed to facility attendance), the effects of distance on utilization were less pronounced.

This distinction will, of course, be less important where there are fewer alternative facilities and irrelevant where only one facility per service exists. This was the case in rural Newfoundland, where Girt (1973) undertook a survey of attitudes towards aspects of health care of 1,400 individuals aged 20 or more, living in seven settlements, each served exclusively by a specific cottage hospital for general medical consultations. Although dealing with hypothetical illness behaviour, a number of interesting trends emerged in this rural locality where health care was at the nearest 10 miles and at farthest 35 miles away. It appeared that 'distance has both a positive and a negative effect on behaviour. Individuals are likely to become the more sensitive to the development of disease the farther they live from a physician but those at a distance may be more discouraged about actually consulting than one living nearer because of the additional effort involved' (Girt 1973, pp. 161–163). The probability of consulting at least once for most diseases seemed to increase with distance from a hospital, until at some point it began to decline. With some diseases, however, sensitivity is apparently so low that only a negative distance effect was discernible. This is true of conditions such as febrile and non-febrile common cold and sore throat (Figure 6.5c), whilst for more 'serious' conditions such as benign hypertension (Figure 6.5a) and normal pregnancy (6.5b), distance has both a positive and negative effect on the probability of an individual consulting. The distance at which this effect changes seems different with different illnesses or conditions, as the diagrams illustrate, which could have important implications for locational planning of facilities.

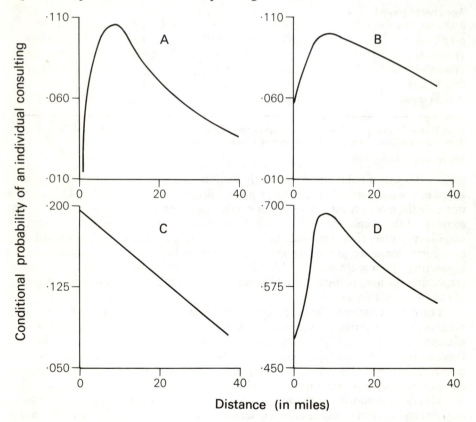

A   Benign hypertension with or without heart disease
B   Normal pregnancy
C   Common cold with or without other symptoms
D   All causes

Figure 6.5   Distance and consultation rates in a rural environment

Source: after Girt (1973)

Comparing the utilization of primary care in urban Hackney (London) and remote Western Isles of Scotland shows marked differences probably caused by availability. Whilst 80 percent of urban residents were within 1 mile (1.6 km) of their GP's surgery, only 14 percent of Island residents were and, whilst 62 percent of the urbanites could reach surgery in less than 10 minutes, 63 percent of Islanders took more than this time to do so. Distance and inaccessibility seemed to have particularly deleterious effects on various types of consultation: only 66 percent of Islanders compared with 74 percent of Hackney respondents had seen a dentist in the previous two years and, for residents over the age of 70, only 23 percent had received chiropody in the Islands compared with 39 percent in Hackney. Therefore, a picture of remoteness and disadvantage was painted both in availability and in some aspects of utilization of primary health services (Williams et al. 1980).

Research into distance decay effects in the utilization of primary care has become more sophisticated by disaggregating respondents by age, sex, education, and mobility. Haynes and Bentham (1982) found that general practitioner consultation rates, amongst other use indices, declined with decreasing access to surgery. However, subgroups of patients seemed to be affected differently. Comparison of the utilization behaviour of adults close to hospital and GP services in Norwich, East Anglia, with that of inhabitants of villages remote from Norwich (and with or without surgeries) revealed important differences. In particular, residents in remoter areas without surgeries seemed to have lower consultation rates. The percentages of respondents attending various types of service in the different locations are shown in Table 6.7, which illustrates the importance of disaggregating the respondents by at least sex, car ownership, and social status. In addition, Haynes and Bentham found differences in utilization patterns between patients with and without long-standing illness, indicating the importance of health status in influencing behaviour. For those with no long-standing illness, Table 6.7 shows that most subgroups contributed to the drop in GP consultation rates from the most accessible to more remote rural areas but with differences being greatest between the young and mobile and the old and immobile.

Serious concern is raised by the finding that groups which probably have a high need for health care, such as the elderly and manual worker households, seem to be using hospital outpatient facilities considerably less than might be expected from their GP consultation rates. Overall, for those with no long-standing illness, there were notably lower consultation rates in remoter villages without a GP's surgery. There was not much difference in referral rates to outpatient facilities by GPs once they had been consulted, except it seemed that GPs in remote rural areas were less likely to refer the elderly, males, and those from manual worker households to an outpatient clinic than were GPs in more accessible rural areas.

This study makes the important point that a simple picture rarely exists in utilization. Therefore, disaggregation of respondents according to specific behavioural characteristics, together with detailed knowledge of local facilities, seems to be essential in this type of 'distance decay' research if facile observations are to be avoided.

Hospitals

It is difficult to generalize about the effects of distance on hospital utilization because this sector of health care is more varied than any other. Hospitals provide locations for 'primary' types of contact, such as the outpatient and emergency services discussed

Table 6.7 *Percentage rates of GP consultations and outpatient attendances by area type and population characteristics (respondents with no long-standing illness), Norwich, England*

Group	GP consultations		Outpatients		
	Accessible rural	Remote rural	Accessible rural	Remote rural	Sample Size
Age: 18–44	22*	9*	16*	5*	313
45–64	23	11	23	11	166
65+	21	27	17	13	93
Sex: Male	18	13	18*	4*	268
Female	25*	12*	18	14	309
Car: Car owning	22*	10*	18*	8*	478
Not car owning	26	20	18	13	100
Social: Manual	22	12	19*	6*	256
Non-manual	21	12	18	12	260

* Significant at the 0.01 level

Source: after Haynes and Bentham (1982)

earlier. They are also locales for major inpatient treatment of acute and chronic conditions. Other hospitals can become homes for mentally ill and mentally and physically handicapped persons. Therefore, any statements will have to be specific to particular sectors of hospital services.

The main focus of this book is on primary care and on general practice in particular. However, much pioneering research in medical geography took place within the North American hospital context, most notably the work of the Chicago Regional Hospital study in the 1960s (De Vise 1968, 1973; Morrill and Earickson 1968a, 1968b; Morrill et al. 1970). This study employed, amongst other techniques, traditional, gravity-type spatial interaction modelling, which it is now accepted offers only limited explanations of functional and spatial relationships. The idealized normative assumptions about the spatial behaviour of both suppliers and consumers restrict its practical utility and inevitably make results less realistic (Thomas 1976; Phillips 1981a; Stimson 1981; Giggs 1983a). However, this normative research did indicate that the use of hospitals varied according to a number of factors specific to institutions, such as amount and volume of service provided, the character of the service area and relative location of the hospital, length of stay, service quality, and age of hospital (Morrill and Earickson 1968a). Other crucial factors governing utilization appeared to be distance, race, religion, and income (Morrill et al. 1970).

More recently, Walmsley (1978) has investigated the influence of distance on hospital usage in rural New South Wales. The use of a general health care facility was found to be at least partly a function of location, in that the chances of admission diminished the further a patient lived from hospital. Distances were considerable in this rural study and some patients whose records served as data sources were travelling up to 40–50 km to hospital. Eighteen small hamlets were sending patients to Coffs

Harbour and District Hospital, the single public hospital in the district, which patients had to patronize or else forgo public health care. Correlations between per capita usage and distance were -0.52 for outpatients and -0.68 for inpatients, significant at the 95 percent and 99 percent confidence levels respectively. At 20 km from hospital, inpatient patronage declined to 40 percent of what it was immediately around the hospital, whilst outpatient attendance fell even further to some 25 percent (Figure 6.6). This suggests that any heightened sensitivity to illness found in other rural research, such as by Girt (1973), does not seem to be reflected in behaviour.

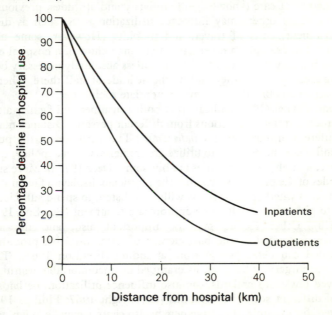

Figure 6.6   Distance decay in the utilization of hospital services in New South Wales

Source: after Walmsley (1978)

The clear message of this paper is that distance decay is evident in the travel patterns of patients and that spatial injustice can arise because people distant from hospital have more restricted health care choices (Walmsley 1978). The wider study on which this paper is based (Walmsley and McPhail 1976) provided details of differences between routine and non-routine demand and by various types of complaint, age and sex of patient, and seasonal factors. It is an instructive and well-conducted report and provides a useful example of empirical research in this field.

Factors underlying differential utilization and distance decay

Distance decay in utilization rates manifests itself for many medical services. However, it has been noted that the effects of distance seem to be different for different groups of people, for different services, and for various ailments. Indeed, it may be questioned whether distance (or accessibility) itself is the major influence or just an 'intervening obstacle' to health care. An important observation is that more research is necessary in this topic to disaggregate users (and to determine *non*-users) according to a variety of behavioural characteristics including those noted earlier. In the use of ante-natal clinics, for example, Kaliszer and Kidd (1981) found that employment status, parity, and age were more important influences than accessibility, although they felt that greater distances than they examined might play a bigger role.

What other factors seem to influence utilization rates and distance decay in health facilities? The list is very long and it is only possible here to suggest some of the more important. It is probably vital to realize that some factors will influence choice or selection of *sources* of care (knowledge, familiarity and attitudes, previous residential patterns, etc.) whilst others may influence utilization *rates*, such as distance from source of treatment, costs of travel, and mobility. However, some might act to influence both. For example, a potential patient may know of a hospital or clinic in a given location but may be reluctant to attend unless absolutely necessary because he or she is discouraged by the journey and by the attitude of staff there. Therefore, some factors may act in combination and may potentiate each other.

Joseph and Poyner (1982) indicate that both consumer and facility attributes will interact to produce different reactions from different persons. Distance may as a result have very different implications for *individuals*. The attributes of the public service which may influence the decision to utilize are numerous: intake policy, quality, type, capacity, price, and physical amenities are but some (Dear 1977a). More subtle factors are the attitudes of the providers: whether the physician displays a favourable 'affective behaviour', that is whether he or she is willing to listen, to spend sufficient time with patients, or to give what is felt to be appropriate treatment (Ben-Sira 1976; Phillips 1979b, 1981a). Other factors may be important also, and the activities of 'administrative personnel' or the bureaucracy of 'administrative procedures' before care is provided can deter people from attending (Freidson 1961). The physical attributes of the surgery or hospital, its cramped or uncomfortable waiting conditions or its perceived physical standards, can also influence utilization, perhaps differently for people of different social backgrounds (Cartwright 1967; Phillips 1981a). It has been suggested, for example, that large new health centres may be 'alien' to some low-status groups and may deter those groups most in need of care (Stacey 1977)

The potential user's characteristics may well affect his choice of facility and frequency of use. Previously, social status, sex, age, and ethnicity were all implicated as possible causes. Education, although closely allied to status/income variables, is also important. The information which a person has will be influenced by variables such as age, sex, friends, and location. A person's education will, in part, govern how such information is used (Joseph and Poyner 1982). Personality may be an equally important variable, but one which is rarely included in social or geographical research and one which may not be satisfactorily measurable (Irving 1975). However, prior knowledge of services and of the availability of care does seem to be important in influencing utilization and there is little doubt that this varies considerably with non-geographic factors such as social status (Stimson and Cleland 1975; Cockerham 1978; Giggs 1983a).

Part of knowledge is that information is gained through experience and this seems to exert a considerable influence on *which* services patients use, if not as obviously influencing *how often* facilities are used. In West Glamorgan, South Wales, Phillips (1979a, 1981a) found that a large proportion of apparently aberrant utilization behaviour could be explained by looking at the past residential history of respondents and their acquired experience. Fairly large proportions of respondents in this intraurban study were travelling beyond their nearest GP's surgery, sometimes going over two or three miles when much closer surgeries were available. Comparison of the current place of surgery attendance with respondents' previous residential areas often explained these ostensibly illogical patterns of travel. For example, in one survey site well endowed with GP surgeries, over 20 percent of respondents were using surgeries outside their immediate home area. However, 75 percent of these persons were actually attending surgeries near to their previous homes and their current attendance patterns were 'hangovers' of their previous attendance.

These were called 'relict patterns of travel' and accord with the idea expressed by some researchers that consumer behaviour after migration is something of a reassessment process of the advantages and disadvantages of maintaining links with old, familiar services (Lloyd 1977). Affiliations with low-order services tend to be broken first, whilst links with personal services such as doctors may be maintained for long periods (Phillips 1979a, 1981a). Social status did influence these relict patterns of travel, partly because the higher social status groups tend to move more frequently and further (Herbert 1972; Herbert and Thomas 1982) and would hence be more likely to have needed to change doctor (Phillips 1979a). However, some other considerations seemed to permit the maintenance of long-distance journeys for medical care. In particular, personal mobility can be important. Car ownership rates are almost always higher for the more well-to-do (although 'car availability rates' are more important for attending day-time services). However, personal mobility can also refer to wider concepts in medical geographical terms, encompassing physical ability to walk to and attend surgery rather than necessitating home visits by physicians where these are provided. Consultations may decrease as distance increases but the assessment by individuals of how easy they find the journey for medical care can indicate, surprisingly, that those who find the journey the least convenient tend to be the most frequent users. The explanation is probably that people who suffer from ill-health both consult most frequently and are liable to find the journey for health care difficult (Ritchie et al. 1981).

Personal mobility in terms of private transport availability is certainly important, particularly in rural areas, in enabling people to travel more easily to health care facilities. In developing countries, however, physical distance and lack of transport can be very real obstacles to the receipt of health care, particularly when this is provided to scattered populations from clinics or 'field stations' at central points. This has been a problem both for providers and users of health services and has particularly hindered the extension of preventive medicine in many developing countries (Orubuloye and Oyeneye 1982). This is in part the reason for the attempted decentralization of medical services by such countries, outlined in Chapter 2, which aims to reduce the 'friction of distance' caused by poor transport availability and low levels of personal mobility. This will not be an easy task, however, for even in many developed countries, accessibility to services and facilities remains the 'rural challenge' (Moseley 1979).

Disadvantaged consumers

The earlier sections of this chapter and the immediately preceding discussion have implied that some sections of communities are less well able to use health services than others. This notion has been developed in a number of other spheres of consumer behaviour and the existence of *'disadvantaged consumers'* is now clearly recognized. These may be disadvantaged on a number of indices and 'the general perspective is that particular sub-groups, such as the lowest social classes or ethnic minorities, are restricted to whatever services exist locally due to a combination of low income and restrictions on personal mobility' (Herbert and Thomas 1982, p. 258). This is well illustrated by reference to car ownership, which considerably eases attendance at medical and other services. In many middle- and upper-status residential districts in Britain, for example, car ownership levels exceed 85 percent whilst in low-status neighbourhoods they rarely exceed 40 percent. Age and sex variations in driving licence holding and restricted car availability often exacerbate these differences and, especially in rural areas, authors have identified some individuals as the 'transport poor' (Wibberley 1978; Phillips and Williams 1984).

Other studies emphasize the constraints on the consumer behaviour of ethnic groups, as noted earlier. A pattern emerges of small, inefficient services, low in quality and sometimes high in price, which can pervade the generally dilapidated ghetto areas of some British and many North American cities (Rose 1971; Herbert and Thomas 1982). The opportunities for obtaining good-quality, reasonably priced medical care can be very few for 'disadvantaged consumers' in many national settings and this must inevitably reduce utilization. In the early 1970s, for example, it is claimed that America's black ghettos had but one-tenth of the physicians they needed, compounding many other aspects of deprivation (De Vise 1973). In the late 1960s, the black population was some 14 percent of the total US population but only 2 percent of America's physicians were black (Shannon and Dever 1974). This is in spite of the fact that racial differences in health are very evident in the USA and elsewhere, the health status and life expectancy of ethnic minorities being almost always inferior to those of the majority of the population (Cooper et al. 1981). Such imbalances in ethnic health and ethnic health care form but one aspect of this at present relatively underresearched area of consumer deprivation in medical geography.

Patients' attitudes

The attitudes of patients and potential patients to their physicians and health providers can also influence attendance rates. If the physician's 'affective behaviour' is unacceptable, perhaps utilization will not occur or will be reduced (Ben-Sira 1976; Phillips 1979b). The patient may have greater empathy for certain physicians, particularly if they are more sympathetic and willing to spend time, as the 'personal doctor' will do (Fox 1960, 1962; McCormick 1976). The whole subject of communications between patients and doctors is vastly complex but important. It has been found that hospital doctors, frequently technology-oriented, often depersonalize and underinform patients (Cartwright 1967; Reynolds 1978; Lawson 1980). Poor communication can be exacerbated by linguistic or cultural barriers but, interestingly, communications seem to be good in many aspects of primary care although there do seem sometimes to be poorer attitudes to some aspects of GP services on the part of lower-status patients, suggestive of a social inverse care law, similar to that existing in the availability of services (Hart 1971; Phillips 1979b).

It is possible to envisage the use of services as a type of trade-off in which the disadvantages of travelling further to use a preferred service outlet are weighed against the problems of not obtaining the service at all or of using a (perceived) less attractive service. This can be researched by questionnaires but revealed behaviour, resulting from a composite of experience and information, is difficult to relate directly to preferences. Another type of trade-off analysis has been performed by Smit and Joseph (1982) in which a game format can be used to identify the preferences of residents for alternative strategies of service provision, given a limited budget for the supply of services. If attitudes to various services could then be related to subsequent utilization behaviour in an explanatory sense, this would be an important development.

The whole area of behavioural research touched on in this chapter is, however, problematic because almost all surveys can be subject to various types of error and bias in data collection and analysis. There are also particular problems regarding the reliability of people's recall of medical service attendances which can sometimes be more than 50% underreported after 12 months have elapsed (Stimson 1983). When attitudes to services are introduced, the explanation of utilization patterns becomes yet more complex.

A conceptual framework for the empirical investigation of health services utilization?

In an ideal world, the needs of individuals for medical services (or other public services) will be satisfied through the use of a convenient facility. It has been evident, however, that many factors can influence the transformation of need into utilization and some have been discussed in this chapter. It is possible to summarize the apparently more important of these diagramatically although it should be emphasized that the focus here is upon *recognized* need rather than other types which were identified in Chapter 3 (Bradshaw 1972).

Model 1 in Figure 6.7 follows the conventional aggregate lines of explanation. Variations in utilization are, it suggests, caused by broad differences in consumer characteristics (such as age, sex, income, and personal mobility) and by the influence of facility attributes (for example, size, cost, quality, and location). As such, it follows quite closely Huff's (1960) representation of the consumer decision-making process. Model 2 offers an alternative level of explanation, taking into account to a much greater extent the variations in individuals' perceptions of a service, of facilities, and of their accessibility. An individual will use a service only when it is perceived that the recognized need should be satisfied, that the facility provides the required and desired service, and that it is convenient. This model has intuitive appeal, based as it is upon consumer behaviour research, with 'action space' proposed as an important bridge between the two conceptualizations (models). Action space is that area of the urban or rural environment which an individual is most familiar with, has knowledge of, and which therefore has an associated perceptual value. Its extent seems to vary with age, sex, occupation and income (because they influence mobility), and activity patterns. It has received increasing attention since Huff's (1960) work on consumer travel behaviour (Horton and Reynolds 1969, 1971; Joseph and Poyner 1982). Action space will probably influence the amount and quality of information available to individuals and, as this will affect perceptions about alternatives and convenience, it can provide the link between the two models in Figure 6.7.

As Joseph and Poyner (1982) point out, it is difficult in practice to separate direct links from those influenced via action space. Indeed, Horton and Reynolds did not attempt to distinguish whether an individual's behaviour was determined by his action

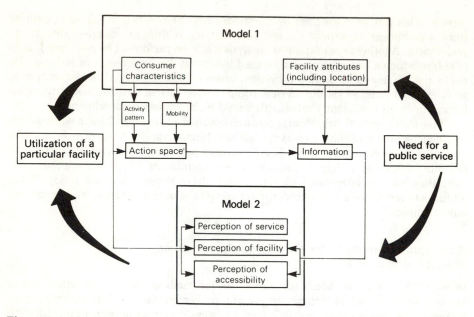

Figure 6.7   A conceptual framework for the investigation of the use of public services

Source: Joseph and Poyner (1981)

space or whether his behaviour created his action space (Thomas 1976). Nevertheless, the relevance of action space to an understanding of consumer behaviour is clear and, in a study of the use of medical, dental and library facilities in Erin Township, southern Ontario, action space did seem partly to explain the location of facilities used. Action space was measured by combining data on activity patterns and mobility for 266 households and it apparently increased with age and social status and was more extensive for men. Knowledge of facilities seemed to be significantly associated with action space but more extensive action spaces were associated with less information about local facilities. As a result, those with more restricted action spaces tended to use local facilities whilst those with a more extensive action space often used alternative facilities (Table 6.8). Whilst recognizing the limits of one case study, this underlines

Table 6.8   *Extent of action space of respondents using local facilities in Erin township*

Action space	Medical		Dental	
	User	Non-User	User	Non-User
Limited	29	35	16	48
Intermediate	41	129	20	150
Larger	1	31	2	30
(Significant at 0.05 level)	$\chi^2 = 21.00$		$\chi^2 = 8.57$	

Source: after Joseph and Poyner (1982)

the need for further research into the nature of the interplay between location of facility relative to knowledge, information, and action space, and suggests that these may be crucial in determining levels of utilization as well as place of attendance. This is perhaps a fair assessment of current levels of understanding of medical services utilization and particularly of the influence of spatial factors on it.

7 Jarvis' Law and the Utilization of Mental Health Care

At several points in this book we have taken pains to point out that health services are complex, often delivered via hierarchical systems characterized by overlapping organizational and spatial structures. It is necessary to acknowledge now that these discussions were deficient in one important respect, namely the failure to point out the existence of a distinctive subset of health services, a subset which will be referred to here as 'mental health care'.

There are a number of sound reasons for considering the use of mental health care facilities separately from that of facilities supplying other forms of health care. In pragmatic terms, the desirability of doing so is suggested by the tradition in most countries of separating the hospital-based treatment of the mentally ill from that of the physically sick. (This tradition and its erosion in recent decades will be discussed in more detail later in this chapter.) Of more fundamental importance in legitimizing the mental health care–physical health care dichotomy is the nature of mental illness and, consequently, its implication for the process by which need for a service is translated into use of a facility offering suitable treatment.

Almost by definition, mental illness implies an impairment of an individual's ability to look after his or her own interests. Dear (1977a) suggests that mentally ill persons may be able neither to recognize the existence of sickness nor (if it is recognized) rationally to evaluate treatment alternatives. This places such persons at the mercy of the health care system, and particularly the organizational infrastructure (the 'referral system') that guides individuals towards appropriate treatment. The imperative for efficiency on the part of referral agents (those who act on behalf of the mentally ill) is furnished in the belief that treatment of mental illness produces benefits to society generally as well as those that accrue to individuals.

The relevance of research, medical or otherwise, on mental illness stems from its extent in modern society. In Canada, for instance, it was estimated in 1965 that 12.7% of all males and 11.7% of all females would at some time in their lives require at least one admission to a psychiatric institution (Dominion Bureau of Statistics 1968). Although spatial variations in the incidence of various types of mental illnesses have long been recognized (Faris and Dunham 1939; Giggs 1973, 1979, 1983b), the spatial aspects of mental health *care* have only recently caught the interest of geographers (Dear 1977a). The stigma and mystique attached to mental illness and its treatment, by the public and by the majority of health care professionals alike, has fostered something of a general neglect by social scientists. More specifically, geographers interested in questions of access and utilization have repeatedly come up against barriers to empirical research erected in the cause of patient confidentiality.

Following a discussion of the evolution/transformation of mental health care delivery systems in Western societies during the course of this century, the remainder of this chapter is devoted to a review of research on revealed access to mental health care services; more specifically, on the role of distance in 'explaining' utilization patterns. Two related studies of mental health care utilization in central Ontario, Canada, will be used to illustrate the nature and extent of geographic effects.

The delivery of mental health care services

In the first half of this century, the systems for the provision of mental health care were relatively simple in most Western countries. Given that most of those suffering from mental illness were not capable of recognizing their illness and/or reacting to it rationally (or at least that this was accepted to be so), need for service was usually determined on behalf of those affected.

In most cases it was the family physician who decided whether a complaint could be treated within the home and family or necessitated, instead, some special environment. For those fortunate enough to afford it, this 'special environment' might be a private nursing home but, for the majority, referral was to the large asylums or hospitals for the insane, the European origins of which are recounted by Rosen (1968). Therefore, potential physical accessibility in the early/mid-twentieth century was largely a function of physical proximity to large institutions for the treatment of mental illness, as modified by the organizational/referral system. Since the mid-point of the century, however, the emphasis upon institutionalizing the mentally ill has declined steadily. Today mental health care is one of the main sectors in which 'deinstitutionalization' is actively developing.

Significant dissatisfaction with the performance of asylums and related institutions for the care of the mentally ill existed in medical and lay circles in the late nineteenth century in most countries, and increased in the twentieth century as the continued overcrowding of asylums reinforced their custodial (as opposed to curative) role (Rothman 1980). However, despite the repeated calls for reform, the monolithic, custodial, and forbidding asylum remained the cornerstone of the mental health care system of 'civilized' countries to the middle of this century.

National reactions to the problems of treating the mentally ill within the asylum framework varied considerably, but reforms were universally facilitated by changing attitudes to, and awareness of, mental illness, and a simultaneous improvement in the ability (through drugs and therapy) to treat such illness (Williams and Lutterbach 1976). Up to and including the early 1960s, deinstitutionalization in both Britain and America was little more than a recognition by psychiatrists and planners that the *long-term* population of mental hospitals was falling, in spite of rising admission rates. This was hazy and not really incorporated into policy (Bennett 1979). However, in the United States, the retreat from the asylum was heralded by the 1955 Joint Congressional Commission on Mental Illness and Health, which advocated the development of services on a local basis (Dear and Taylor 1982). Following the Community Mental Health Centers Act of 1963, the United States government has, through its various agencies, promoted the 'localization' of mental health care, the central feature of which has been the Community Mental Health Center (CMHC). In consequence of this deinstitutionalization policy, the population of state and county mental hospitals fell from a peak of 559,000 in 1955 to 193,000 in 1975, despite the fact that prior to 1955 the number of inmates had been increasing at the rate of 2 percent per annum (Dear and Taylor 1982). These data reflect the enormity of the task transferred from the asylums to the new locations of care.

In contrast to the United States, deinstitutionalization in other countries, in the United Kingdom and Canada for instance, has been less structured but nevertheless evident. In England and Wales, the resident population of mental hospitals fell by 28 percent between 1951 and 1970 (Scull 1978). However, it was not until the Hospital

Plan for England and Wales of 1962 that the administrative and general responsibilities of local government authorities to provide services for the mentally ill were outlined. This followed the report of a commission on mental illness in 1961, *Action for Mental Health*, which stressed the need to avoid the debilitating effects of institutionalization as much as possible, the benefits of early return of patients to the community, and the need to maintain discharged patients in the community (Bennett 1979).

In Canada, each province has pursued its own policy, with Saskatchewan at the forefront of the deinstitutionalization movement (D'Arcy 1976). In all provinces, the emphasis has been upon the transfer of responsibility for treatment from the provincial mental hospital to psychiatric inpatient and outpatient units of general hospitals or to community-based public health or social service agencies, both public and charitable (Williams and Lutterbach 1976). In Ontario, the net result of deinstitutionalization was to reduce the number of residents of provincial asylums by about 75 percent between 1960 and 1976!

The substantial and rapid change in the structure of mental health care delivery systems consequent upon the adoption of deinstitutionalization policies has been instrumental in a radical reshaping of accessibility issues. In broad terms, potential access will still be determined by the availability of services. Indeed, the adequacy of community care facilities in some jurisdictions has been seriously questioned. Commentators on both sides of the Atlantic have suggested that mental hospitals were more enthusiastic in discharging patients into the community than were community authorities in providing suitable residential and treatment centres (Clare 1976; Bassuk and Gerson 1978).

Notwithstanding the importance of the availability issue, the most dramatic implications of deinstitutionalization for accessibility to mental health care services stem from the increasing complexity of services and delivering facilities that has accompanied the 'localization' of mental health care. On the one hand, the move from a few large residential-care institutions to a larger number of small, locally based residential or outpatient treatment centres has undoubtedly given rise to real improvements in the geographical accessibility of care, in that the average distance between facilities and potential clients has invariably been reduced. On the other hand, the increasing complexity of the overall system – and the often necessary specialization of constituent facilities – has the potential for substantially increasing organizational or informational barriers to utilization. Therefore, there is a real possibility that the gains in terms of proximity consequent upon decentralization might, at least in part, be offset by losses in terms of broader accessibility resulting from increased organizational complexity. Given the organizational structure of health care delivery (Chapter 3), the extent of this trade-off depends largely upon the effectiveness of referral/directing procedures.

Prior to discussing the degree of distance decay evident in patterns of revealed accessibility to mental health care facilities, it is worth while acknowledging the existence of an alternative research thrust (by geographers) on the community health care question. In this book, proximity to facility has invariably been viewed in a positive light; 'nearness' is equated to potentially greater physical accessibility. There is, however, a tradition in geography of viewing public facilities (and those associated with community mental health care in particular) in a dualistic manner. To user groups, proximity to facilities is nearly always considered to be beneficial, while to non-user groups it might often be viewed as harmful. This appearance of a Jekyll persona to clients and a Hyde persona to (non-user) neighbourhood residents has been noted and extensively discussed by Dear (1977a, 1977b), who distinguishes between 'location as access' and 'location as externality'. Incomplete though it may be, evidence in North America, where deinstitutionalization of mental health care has advanced furthest, suggests that externality effects will often override access consideration (and 'milieu' considerations) in the determination of locations for community care facilities. Dear (1977a), for instance, suggests the concentration of such facilities in inner-city neighbourhoods of North American cities to be strong evidence for the strength of (suburban) neighbourhood opposition to mental health care facilities based upon their supposed negative externalities. This situation serves, of course, to limit the presumed benefits, in terms of increased proximity, of decentralizing mental health care into the community. Readers interested in these externality issues are directed to Dear and Taylor (1982), who document the most rigorous study to date (sited in Toronto, Canada) of attitudes toward mental illness, mental health care, and mental health care facilities. In Britain, for example, public awareness of the deinstitutionalization process often only emerges when specific communities attempt to resist the location of facilities such as hostels for discharged mentally ill persons within their boundaries.

Jarvis' Law

Geographers have frequently observed that human behaviour is responsive to locational relationships; indeed, distance decay in interaction remains a basic tenet of the spatial organization theme in geography. Distance decay in utilization patterns has been observed for public facilities in general (Massam 1975) and for health care facilities in particular (Shannon and Dever 1974). However, although the influence of location on the utilization of mental health has been widely conceded, its specific role remains ambiguous (Dear 1978). Here the intention is to remove as much of this ambiguity as possible.

Distance decay in utilization is taken here to mean that rates of utilization of a particular facility (for example, per 10,000 residents) will be inversely related to distance from that facility. This relationship was first reported in 1851, by Jarvis, who stated that 'the people in the vicinity of lunatic hospitals send more patients to them than those at a greater distance' (Shannon and Dever 1974, p. 111). Since that time, this distance decay effect has been referred to as 'Jarvis' Law', and is one of the 'classic' geographical statements on the friction of distance effect.

Any assessment of Jarvis' Law should, however, be set within a realistic framework. First, it must be kept in mind that locational relationships are in no way the sole determinant of utilization (Chapter 6). It has further been suggested that three broad groups of factors are at work in determining mental health care utilization patterns: the characteristics of the *service* being provided; the characteristics of the *client* population; and the *location* of the service. In terms of the service being provided, utilization might be influenced by intake policies, quality and type of service offered, size and capacity of facility, and price of service (Dear 1976). In connection with the client population, the fundamental variables are those related to the incidence of mental illness, regardless of whether they be genetic or environmental in nature. The basic locational variable is proximity (that is, distance from place of residence to facility), although it should be acknowledged that, as Dear (1976) and others, such as

Tischler et al. (1972), have suggested, non-proximity factors like catchment policies and referral practices have obvious locational implications.

Given the complexity and known interlocking of factors influencing utilization, the identification of the specific role of the main locational variable, 'distance from facility', is clearly a challenging task. Two strategies for evaluating the role of distance are possible, the individual/behavioural approach and the aggregate/ecological approach.

The behavioural underpinnings of distance decay effects in mental health care utilization patterns are only partially understood, and poorly appreciated (Smith 1976). In general terms, for an individual, the failure of need for a service to be translated into use of a facility results from three possible causes: (1) need is not recognized; (2) need is recognized but there is ignorance of opportunities for treatment; or (3) need is recognized and opportunities for treatment are known *but* there is a reluctance to travel to obtain treatment. Shannon and Dever (1974) suggest that a willingness to care for the mentally ill in the home or 'community' may well be a concomitant of this reluctance to undertake travel. If information flows display distance decay (as there is very strong evidence that they do) and reluctance to undertake travel is positively correlated with distance, propositions (2) and (3) are both plausible behavioural explanations for distance decay in utilization rates. Moreover, if the recognition of need is strongly dependent upon professional (medical) intervention and the source of this intervention is located *at* the facility in question, proposition (1) is also a contender!

Unfortunately, primarily as a consequence of the difficulty of access to individuals in need of care (and difficulty in obtaining reliable responses from individuals who may be irrational), no rigorous studies of utilization at the behavioural level exist, at least to our knowledge. Realistically, opportunities for empirically evaluating the role of distance in the determination of patterns of utilization exist only at the aggregate/ecological level.

Ecological case studies have addressed two major questions relating to the role of distance: first, whether there is a 'substantial' (however defined) distance decay effect in admission rates; and, second, how important distance-from-facility is relative to other variables in the determination of admission rates. Although there exists some evidence that individuals will often purposefully make apparently irrational spatial choices – for example, Hankoff et al. (1971) and Miller (1974) noted patients by-passing convenient facilities to obtain some sort of anonymity – 'substantial' distance decay in the utilization rates of mental health care facilities has been reported in several diverse situations. Examples include Sohler and Thompson's (1970) study of distance decay effects in the utilization of state mental hospitals in Connecticut, and Davey and Giles' (1979) analysis of admissions to a mental hospital in Tasmania. However, distance decay effects appear to have been weak in others (Miller 1974; Dear 1976; Sankar and Mintus 1978; Goodman and Siegel 1978). The reasons for lack of consensus concerning the prevalence and strength of distance decay effects are best considered in connection with the relative importance of distance (versus other variables) in determining patterns of utilization.

A review of studies of mental health care utilization reveals a marked relationship between the apparent importance of the distance variable and the scale of analysis, such that in 'regional' studies distance effects are invariably revealed to be more important than in studies carried out at lower levels of spatial aggregation. This suggests two possibilities. First, it might be that individuals are insensitive to small increments of distance and, by implication, that behavioural responses are only observable for large 'jumps' in distance. Second, it might be the case that behavioural adjustments to *small* increments in distance are subsumed within, or masked by, greater variation related to other, non-spatial variables. Studies such as those reported by Sohler and Thompson (1970) for Connecticut and Davey and Giles (1979) for Tasmania provide unambiguous evidence for the existence of distance decay effects in the utilization of large 'regional' institutions. By contrast, one may consider a study in (metropolitan) Hamilton, Ontario, reported by Dear (1976, 1978) as illustrative of the problems encountered at a submetropolitan scale.

Using an ecological framework, Dear analysed the utilization of 11 mental health care facilities in the Hamilton area (ranging from community mental health centres to a psychiatric hospital). For all but two of the facilities, significant distance decay curves were fitted (Table 7.1). The level of explanation (r^2) attached to the curves suggests a strong correlation between distance from facility and utilization of *that* facility. However, on considering the relationship between these same utilization rates and a set of service- and client-related variables, Dear found two client variables, 'percentage of single adults' and 'median income', to be the most significant predictors of utilization rates for most facilities. In this multivariate model, distance-to-facility was statistically significant for only 2 of the 11 facilities. Dear's (1976, 1978) analysis demonstrates that

Table 7.1 *Regression analysis of mental health facility utilization as a function of client distance from facility: Hamilton, Ontario*

Facility Type	Best Least Squares fit (Type of Curve)	Coefficient	r^2	t statistic
OUTPATIENT				
Unit 1	Hyperbolic[a]	2.75	0.042	1.52
Unit 2	Power[b]	−0.64	0.233	−3.02**
Unit 3	Hyperbolic	6.07	0.590	3.98*
Unit 4	Hyperbolic	9.71	0.458	6.24**
Unit 5	Power	−0.50	0.130	−2.61*
Unit 6	Power	−0.90	0.453	−3.86**
INPATIENT				
Unit 1	Logarithmic[c]	−3.22	0.667	−5.86**
Unit 2	Power	−0.64	0.239	−3.54**
Unit 3	Power	−0.56	0.244	−2.32*
Unit 4	Power	−0.83	0.544	−4.33*
PSYCHIATRIC HOSPITAL				
Unit 1	Logarithmic	−0.11	−0.16	−0.78

a $Y = A + B/X$
b $Y = AX^B$
c $Y = A + B \log (X)$
 * Significant at the 0.05 level
** Significant at the 0.01 level

Source: Dear (1978)

the role of distance factors is difficult to separate from that of other variables at the intraurban, submetropolitan scale. This difficulty may arise because of conceptual overlap (for example, poor people are reluctant to undertake expensive travel, so the differential distribution of the poor across the distance range from a facility may distort the distance decay effect) and/or because of the problems of reliably estimating relationships from ecological data using regression (see Chapter 4, 'Factors influencing physician location').

The major conclusion that can be drawn from the above discussion is that the debate over Jarvis' Law has been obscured by the mixing of results obtained through different research designs. On the whole, most empirical tests using ecological data have been crude (often by necessity rather than design) in the sense that they have paid little attention to the complex reality of the mental health care utilization system. Two of the most important features of this reality, the role of referral systems and of diagnosis, will provide the focus for the remainder of this chapter. Appreciation of their role is fundamental to a proper appreciation of the influence of facility location patterns on utilization, particularly in the context of deinstitutionalized and decentralized supply systems.

Referral as a modifier of distance decay in revealed accessibility

The potential role of the organization of the supplier in modifying accessibility to health care has been stressed at numerous points in this book. Nowhere is this potential greater, of course, than in connection with the use of *mental* health care facilities. In spite of changes in treatment philosophies and institutional organization throughout this century, the physician continues to play a pivotal role in the utilization process.

Given that the mentally ill are, by definition, irrational to some extent, it often falls upon others to make decisions for them. This task is often undertaken by the spouse, relatives, or friends, but the family doctor invariably looms large in the *medical* decision. The critical point here is not who these intermediaries are (although this is clearly important too) but that these 'referral agents' exist. Their existence means that an understanding of distance decay in utilization rates will require an appreciation of the attitudes to a service and to use of a particular facility on the parts of referral agents as well as of those in need of care. It is the referral agent as well as the individual needing care who may weigh the pecuniary and time costs of travelling for treatment against the potential costs of not undertaking treatment.

Geographical studies of mental health care utilization have rarely been able to evaluate the role of referral or even to characterize patterns of referral, usually because of the nature of their data bases. Therefore, a case study (Joseph 1979) of the utilization of outpatient mental health care services in central Ontario, Canada, is worth recounting in some detail. Peterborough mental health catchment area had a 1976 population of nearly 220,000, with 60,000 living in Peterborough itself, the main urban centre. A full range of inpatient and outpatient psychiatric services was provided to the area by the Peterborough Civic Hospital whilst several other hospitals in the region provided limited bed space with certain clinics providing restricted outpatient services (see Figure 7.4).

Data were obtained on the age, sex, place of first appointment, residential location by rural township or urban census tract, and referral source for all new outpatients registered in the Peterborough mental health catchment area in 1976, 883 persons in all. Referrals were placed in five categories: those by (1) psychiatrists; (2) general practitioners; (3) other medical personnel (e.g. public health nurses); (4) non-professionals (e.g. self or a relative); and (5) institutions (e.g. schools).

Distance decay in the pattern of new admissions

The dominant position of the Peterborough facility in the delivery system is evident in that 767 of the 883 subjects in the study made their first appointment for treatment at the central facility as opposed to the peripheral clinics. Moreover, because of the limited services available at the clinics, the majority of the remainder were required to obtain most of their treatment at the central facility in Peterborough. Therefore, in 1976 the system was essentially based on a single facility; access to outpatient mental health care services within the catchment area was dependent primarily upon access to the Peterborough Civic Hospital. This greatly simplified the assessment of distance decay effects.

The relationship between the number of new patients per capita in 1976 and distance from the township or census tract centroid to the central facility is manifestly imperfect (Figure 7.1), while the best fit regression has a coefficient of determination

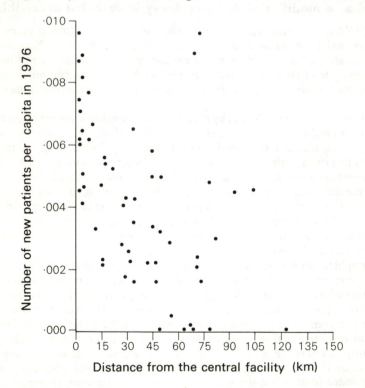

Figure 7.1   The influence of distance from the main psychiatric unit on the number of new patients per capita, Peterborough, Ontario, 1976

Source: Joseph (1979)

(r^2) of only 0.2386. The weakness of the distance decay effect points towards the importance of other variables such as the incidence of mental illness and mental problems and/or the existence of other, non-geographical barriers to utilization. Because data were not available on the incidence of mental illness and mental problems in the study, or on non-geographical barriers to utilization, resort was made to a less satisfactory but practical alternative, a multiple correlation and regression analysis of utilization rates on a set of demographic and socio-economic variables (Table 7.2). Although attempts have been made to interpret such variables as surrogates for illness incidence or for various forms of accessibility, in this study they were treated simply as alternative variables to distance.

Table 7.2 *Selected simple correlation coefficients for utilization and demographic and socio-economic variables: Peterborough*

Independent variables	New patients per capita in 1976	Distance to central facility
Family incomes ($\bar{x}$)	0.3452**	−0.6408**
Persons per room ($\bar{x}$)	0.2399*	0.0053
Persons per household ($\bar{x}$)	0.1201	−0.3882**
Under 20 years old (%)	0.1563	−0.3374**
Over 65 years old (%)	−0.5041**	0.6848**
Unmarried (%)	0.2292*	−0.4930**
Less than grade 9 education (%)	−0.4375**	0.5056**
Foreign-born (%)	0.2618*	−0.3892**
UK origins or ancestry (%)	0.0750	0.1217
Roman Catholic (%)	0.3086**	−0.6151**
Distance to central facility (km)	−0.4885**	

* Significant at the 0.05 level
**Significant at the 0.01 level

Source: Joseph (1979)

Consider first a model incorporating 10 demographic and socio-economic variables, but not the distance variable. Excessive multicollinearity (lack of statistical independence) within the set of independent variables makes it difficult to assess accurately their relative importance within the regression model, so attention was focused on the level of explanation (R^2) and the simple correlations (r). The nature of the latter suggests that utilization of mental health care facilities in the study area tended to be higher in populations with a low proportion above retirement age, a relatively high level of both educational achievement and income, and a high proportion of foreign-born residents and Roman Catholics. However, the low level of explanation associated with the multiple regression model, $R^2 = 0.4897$, diminishes the pertinence of these observations and points towards the importance of variables not included in the model, of which distance is obviously one. However, a model expanded to include the distance variable had an R^2 value of only 0.5460, although the existence of statistically significant collinearity between the distance variable and several of the socio-economic and demographic variables means that the 0.0563 improvement in the level of explanation constitutes a measure only of the *minimum importance* of the distance variable (Table 7.2).

In addition to the technical (regression) problems of assessing the importance of a single independent variable (distance) within a collinear data set, correct identification of its role in determining utilization patterns depends also on the applicability of a basic assumption for an analysis of this sort – that the relation between the dependent variable and each of the independent variables is structurally consistent through the study group. If relationships vary systematically in structure among identified subgroups, a model calibrated for the whole data set will not properly identify their real nature and extent. It has already been noted that the referral agent acts as an organizational filter within health care delivery systems, and Joseph (1979) proposed that this filtering effect interacts with location in a systematic manner to modify distance decay effects in utilization.

The role of the referral agent

It was apparent that the effect of location upon utilization differed among the five referral groups when distance to facility was plotted graphically against admission rates for each of them (Figure 7.2). The effect of location on intensity of utilization appeared to be strongest for referral by psychiatrists and weakest for referral by other medical personnel, with the remaining groups spanning the range between them. This could conceivably have resulted either from differences among referral groups in the willingness of individuals to undertake travel for treatment or from the spatial distribution of referral agents. The former possibility seems quite remote, but the latter undoubtedly held true for psychiatrists since they all resided and worked in Peterborough. In this case, however, it was the spatial distribution of general practitioners that was really of relevance because a patient had first to be referred to a psychiatrist by a general practitioner.

Given that published data showed the distribution of general practitioners relative to population to be fairly uniform through the study region (Haliburton, Kawartha and Pine Ridge District Health Council 1977), Joseph (1979) proposed that the observed differences among the five referral groups in the response of intensity of utilization to location reflected either a lack of information about service availability on the part of certain referral agents or a reluctance on their part to dispatch patients for treatment at a distant location. Both possibilities underline the importance of information flows to referral agents, concerning the existence of services and their importance for the well-being of their clients.

The considerable variation in the utilization–distance relation among the five referral groups strongly suggests that the overall effect of location upon utilization could not have been assessed accurately by the regression model. Indeed, an underestimate seems apparent when the distance effect is considered in terms of a core–periphery contrast (Table 7.3). The difference in the rate of overall utilization between the two regions is considerable, and well beyond that which could conceivably be attributable to systematic variation in the incidence of mental problems and mental illness. In addition, the rates for the individual referral categories underline the complex but seemingly systematic role of the referral system in the production of contrasts between the city and surrounding areas.

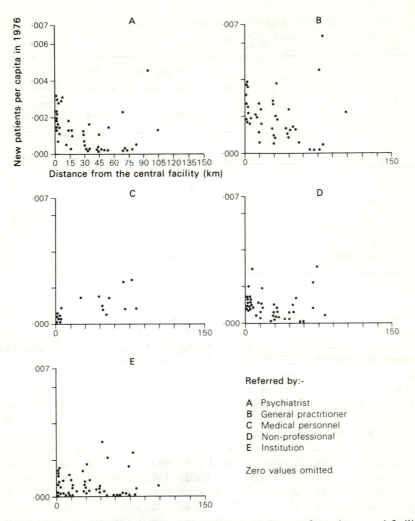

Figure 7.2   The number of new patients per capita and distance from the central facility in Peterborough, by referral agency

Source: Joseph (1979)

The results reported by Joseph (1979) suggest that distance-to-facility is an important factor in the determination of the pattern of utilization of outpatient services. It now remains to question whether a distance decay effect would exist for more serious forms of mental illness.

Diagnosis and distance decay

It has often been proposed that admission rates for more severe mental disorders are more weakly related to distance-from-facility than those for less severe ones (Shannon

Table 7.3 *Peterborough, Ontario: City–surrounding area contrasts for selected utilization rates (core–periphery contrasts)*

Sample	New patients per 1,000 inhabitants		
	City and surrounding area	City	Surrounding area
Total	4.505	6.574	3.633
Referral by psychiatrists	1.082	1.927	0.653
Referral by general practitioners	1.821	2.650	1.472
Referral by other medical personnel	0.362	0.155	0.450
Referral by non-professionals	0.735	1.256	0.515
Referral by institutions	0.556	0.585	0.544

Source: Joseph (1979)

and Dever 1974). For the least severe disorders, usually treatable on an outpatient basis, utilization rates are likely to be strongly affected by distance-sensitive flows of information to potential patients and referral agents and the attitude of both groups to the value of treatment in light of the travel costs and inconvenience. Even for moderately serious complaints, not usually treatable on an outpatient basis, it is possible that families, or even the medical establishment, may be reluctant to place the individual concerned at a distant institution (Shannon and Dever 1974). However, for the worst emotional and behavioural disorders, the severity of illness and the imperative need for treatment are invariably recognized regardless of distance from potential sources of care. However, substantive evidence on the sensitivity of distance decay to differences in diagnosis is limited although one study that has addressed this question complements the referral study outlined above (Joseph and Boeckh 1981). This study combines the outpatient data set (new admissions in 1976) already described with data on the age, sex, residential location by rural township, and diagnosis for all new admissions to inpatient facilities in Ontario originating from the Peterborough catchment area in 1976. The classification of diagnosis is given in Table 7.4.

Assessing the impact of diagnosis on distance decay in admissions rates in data sets of this sort is a challenging task. In addition to systematic distortions in the distance decay effect arising from the variable structure of populations and the complex impact of distance on utilization (resulting, for example, from referral systems), random distortions are introduced into analyses of distance decay in admission rates by the nature of the data used. Researchers commonly seek to infer general relationships in populations from examination of sample data. The reliability of these inferences depends primarily upon the size of the sample. The adequacy of sample size is related to the level of aggregation used in the study, in terms of both utilization rates and space.

Table 7.4   *Near-far contrasts in utilization (rates per 10,000): Peterborough*

		Catchment area			
		Peterborough (pop. = 59,683)		'Remainder' (pop. = 158,177)	
		No.	Rate	No.	Rate
Diagnosis	1	36	6.0	93	5.9
	2	32	5.4	156	9.7
	3	26	4.4	64	4.0
	4	211	35.4	432	27.3
	5	159	26.6	198	12.5
	6	91	15.2	269	17.0
Diagnosis Group	1	94	15.7	313	19.8
	2	461	77.2	899	56.8
	3	382	64.0	501	31.7
Total		937	157.0	1713	108.3

Note: Diagnosis
1 = Schizophrenia
2 = Affective psychoses
3 = Paranoid states and specified and unspecified psychoses
4 = Neuroses
5 = Alcoholism
6 = Transient situational disturbances, personality disorders, cephalagia and specified non-psychotic
     mental disorders
Source: Joseph and Boeckh (1981)

As in many studies of mental health care utilization, the data analysed by Joseph and Boeckh (1981) pertain to admissions over a short time period: in this case, one year. Given that the overall incidence rate of mental illness in the population is relatively low, this had strong implications for the way in which the data could be manipulated. If the study area was divided into too many subareas, or utilization was divided into too many categories, the possibility of obtaining spurious results would be increased. Indeed, in any study, only a very large sample would be sufficient to capture the variation in the population of incidence of precisely defined events, such as single illnesses, over very small areas. However, although the reliability of samples decreases with increasing disaggregation, so, unfortunately, does the insight produced by analysis.

Therefore, in structuring any analysis, the researcher is influenced by the degree of disaggregation that the sample size permits as well as by theory or intuition. In the Joseph and Boeckh study, the choice was between aggregation across space, where a maximum of 45 areas was possible, and aggregation of utilization categories, in which a maximum of 7 diagnoses was available. Both routes were explored. First the maximum disaggregation of space and an aggregation of diagnosis is considered.

Assessment of distance decay effects: aggregation of diagnosis

The inpatient diagnoses, 1 to 6 in Table 7.4, were aggregated into two groups: diagnoses 1 to 3, the more serious diagnoses, into Diagnosis Group 1; and the less serious diagnoses, 4 to 6, into Diagnosis Group 2. The already aggregated outpatient group remained intact as Diagnosis Group 3.

The relationship between utilization rate and distance to Peterborough (the facility locus of the mental health catchment area) is plotted for total utilization and each of the three diagnosis groups in Figure 7.3 (the regression fitted is rate = $a + b1n$ distance). Although the regression for Diagnosis Group 1 is not statistically significant at the .05 level (whereas it is for each of the other groups), the direction of the relationships is consistently negative. The magnitudes of the correlation coefficients (r) indicate distance decay effects in utilization rate to be stronger for the outpatient and less serious inpatient groups (Diagnosis Groups 3 and 2) than for the serious inpatient group (Diagnosis Group 1). Unfortunately, Joseph and Boeckh were unable to estimate the extent to which the unexplained variance in utilization rates $(1 - r^2)$ was attributable to systematic forces not included in the simple model, as opposed to random error resulting from the sample characteristics of the data. However, it is interesting to note that the correlation coefficient between *total* utilization and the log of distance is larger than any one of the coefficients for the individual diagnosis groups. This illustrates how sensitive results such as these are to the level of aggregation in the dependent variable. Given the known variability amongst diagnoses, it is quite probable that the correlation coefficient for total utilization is spurious.

Joseph and Boeckh found that the insight gained through simple correlation and regression analysis was limited by the necessary simplification of the facility system that underlay it (namely, the equating of proximity effects with 'distance from Peterborough'). Therefore, as a complement, they offered a more complete but less easily interpreted impression through the mapping of location quotients for each of the four utilization rates considered in the correlation and regression analysis (Figure 7.4). The location quotient in area i for diagnosis k, Q_{ik}, is given by

$$Q_{ik} = (A_{ik} / P_i) / (\sum_i A_{ik} / \sum_i P_i)$$

where:

A_{ik} = number of admissions in area i for diagnosis k;

P_i = population of area i.

A location quotient in excess of 1.0 indicates a rate of admission greater than that for the catchment area as a whole; one of less than 1.0, a lower than average rate; and a figure of 1.0, a rate at the system norm. The use of a location quotient of this simple form was considered superior to the mapping of utilization rates because it translates the different rates onto the same, easily interpreted scale. The most important limitation associated with the use of location quotients is that only a subjective visual assessment of the different maps is possible (see Chapter 5 for a fuller discussion of location quotients).

In the case of total utilization, the pattern of location quotients appeared to be weakly associated with the facility system. Although three of the areas in which inpatient facilities were located display location quotients in excess of 1.0, the other two do not. Moreover, the pattern of location quotients in areas proximate to the site of inpatient facilities is also unclear; some are high, whereas others are low. This confused picture might well be a consequence of the aggregate utilization rate being used.

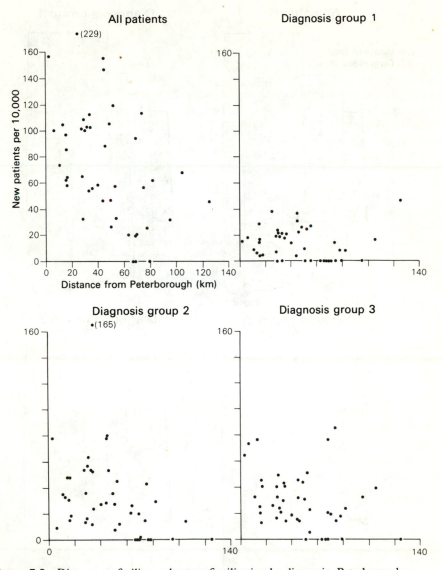

Figure 7.3 Distance to facility and rates of utilization by diagnosis, Peterborough

Total utilization (all patients)   rate = 234.90–44.24 1n distance,
                                                                     $r = -0.4515$

Group 1 (serious diagnoses)   rate = 20.16–1.89 1n distance,
                                                                   $r = -0.1169$

Group 2 (less serious diagnoses)   rate = 84.22–14.51 1n distance,
                                                                            $r = -0.3624$

Group 3 (outpatients)   rate = 71.01–11.42 1n distance,
                                                       $r = -0.3998$

Source: Joseph and Boeckh (1981)

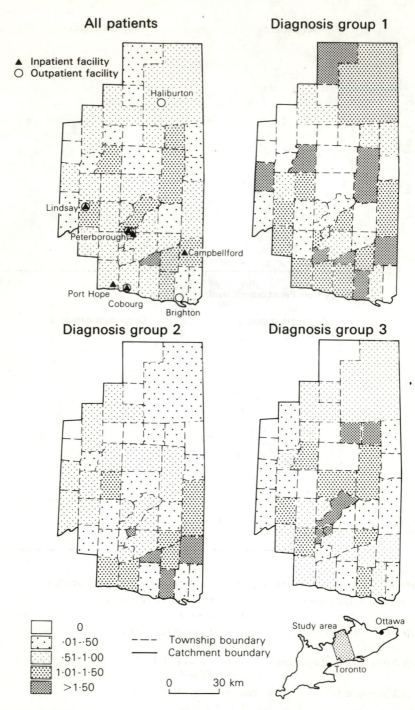

Figure 7.4 Location quotients for the different diagnosis groups (figures above 1.0 indicate higher-than-average admission rates)

Source: after Joseph and Boeckh (1981)

The pattern of location quotients for each of the three diagnosis groups suggests quite different relationships with facility location. The pattern for Diagnosis Group 1 (more seriously ill inpatients) is irregular, which might well reflect the systematic impact of a highly variable incidence rate over space. On the other hand, for Diagnosis Group 2 (less seriously ill inpatients), rates of utilization appear to be more strongly associated with facility location. Indeed, location quotients in excess of 1.0 are recorded for all but one of the areas in which inpatient facilities are sited, and areas proximate to treatment centres generally display higher location quotients than do those further away. For Diagnosis Group 3 (outpatients), the distance decay away from Peterborough is even more marked, reflecting the pre-eminence of this centre within the outpatient treatment system.

Assessment of distance decay effects: aggregation of space

Aggregation of space has often been used to compensate for deficiencies in sample size. Indeed, any study which assigns individuals to areas for the purpose of calculating rates of utilization obviously involves some aggregation of space. (Aggregation of space is interpreted here as aggregation *beyond* the level at which the data are reported). This might involve the aggregation of areal units by their distance from facility and the calculation of zonal rates, or the calculation of some cruder rates, for instance, arbitrarily classifying areas as 'near–far' and assigning them rates. In view of the complexity of the system under study and the relatively small number of reporting areas, Joseph and Boeckh (1981) favoured a near–far comparison rather than the calculation of zonal rates (Table 7.4). These data underline the importance of facility location within this utilization pattern and reinforce the earlier interpretation of the role of diagnosis as an important modifier of distance decay effects. Peterborough, the urban core of the mental health care catchment area, had an overall utilization rate 45 percent higher than that in the remainder of the catchment area. However, when equivalent utilization rates were calculated for each diagnosis and for the three diagnosis groups, it was apparent that this picture was not uniform; near–far differentials were greater for less serious than for more serious diagnoses. A comparison of the differential admission rates for alcoholism (diagnosis 5) and schizophrenia (diagnosis 1) illustrates this well.

Distance decay in admission rates: a retrospect

The Peterborough studies indicate the degree of crudity inherent in Jarvis' Law. Referral patterns and diagnosis are two important modifiers of distance decay effects in admission rates, but there are undoubtedly others. A number of obvious lessons are to be learned from the Peterborough work. It seems that the work of geographers is particularly sensitive to the dangers of spatial and categorical aggregation. Indeed, the chief limitation of any analysis of aggregate utilization behaviour lies in the strength of the assumptions made. Of course, restrictive assumptions are needed to make feasible any analysis of rates as opposed to individual incidence, but it is necessary to keep their implications in mind. This applies to assumptions concerning the level of spatial aggregation, the aggregation of respondent characteristics, and the way in which the service system is assumed to work.

It remains for us to note, yet again, that research on utilization patterns per se, and on their sensitivity to the changing nature of supply systems in particular, is a prerequisite for a better understanding of fundamental questions of availability and equity in access. To evaluate properly the impact of decentralization on accessibility patterns, comprehensive information (across time and space) is needed on the role of locational relationships in determining utilization and on their importance (relative to aspatial factors) within the whole access jigsaw. At present, geographers are some way from this ability, struggling as they are to obtain substantive data on *individual* utilization behaviour. Enough evidence is available, however, to underline the importance of the task.

8   Spatial Aspects of Health Care Planning

This chapter reviews the major aspects of health care planning, particularly as they influence accessibility and utilization of health services. As discussed in Chapter 1, it is important to remember that health care in most societies today still means 'illness' care. Services are generally curative, sometimes preventive, but almost always concerned with medical intervention. As a result, a vast range of activities and sectors which contribute to health and well-being are largely excluded from health planning: environment, housing, education, nutrition, income and so on. It should, therefore, be appreciated that a focus purely on planning policies and their effects on distribution is somewhat restrictive.

The nature of health care planning itself is also somewhat unclear, particularly its spatial aspects. In Chapter 1, epidemiology was introduced as a discipline which has evolved to encompass a similar range of interest to that of medical geography, itself arguably at least partly a form of spatial epidemiology. Most epidemiologists and health statisticians have, to date, interpreted their role in quite limited terms, confining their interest almost exclusively to the determinants and parameters of health, sickness, and disease prevention. Additionally, they were often physically removed from the decision-making process, which was administrative and political. Indeed, there was a restrictive concept in most nations of the proper place a scientific discipline should take in relation to administration; and *health planning* developed in terms of the effectiveness of clinical treatments, and of various drugs and interventions: the classical foci of epidemiology (Alderson 1976; E.G. Knox 1979; Parry 1979). *Health care planning*, on the other hand, is broader and concerns the administration, location and staffing of health care services and their relation to other social services.

The separation of health care planning from health planning as a whole is arbitrary but perhaps desirable: 'its utility can be argued on the grounds that measures for maintaining health, and measures for providing health care, are in many countries administered through different pathways' (E.G. Knox 1979, p. 3). This often implies differing roles for medical geographers. In the USSR, it seems that only the more ecological, disease-oriented medical geography is recognized, whilst health care planning is left to the administrative planning hierarchy outlined in Chapter 2 (Learmonth 1978). In the USA, Sweden, Britain and Canada, on the other hand, there seems to be greater scope and potential for the contributions of academic health care geographers to planning.

It is also possible to overstate the importance of health care planning in effecting improvements in health. The relationships between health service provision and good health have, as suggested in Chapter 1 and elsewhere, sometimes been exaggerated (Howe and Phillips 1983). Of equal importance can be levels of nutrition, education, housing and sanitation. However, it is generally accepted that there is some congruence between health and curative and preventive health services, even if it may be that more ill-health is identified where services are better! That health care services provide at least one element in the support of human well-being is not in debate, so their planning (or non-planning) requires attention.

Health care planning: spatial aspects

It would, of course, be amazing if the efficiency of health services in reaching patients had not long concerned those working in medicine and health administration in *any* country. However, as Chapter 2 emphasized, different political and economic systems produce very different health care delivery systems, and the spatial component in their planning (and hence the potential role of the geographer) varies considerably.

It would be unfair to claim that geographers should have the prime place in health care planning although their skills should certainly provide useful input to a planning team. This has been the case in a number of countries, Sweden providing perhaps the classic example in the early 1960s when a geographer, Godlund, was involved in planning the sites of alternative hospitals when the existing system was being extended. The proportion of people who could be reached under different locational strategies and with limitations on possible patient travel time were calculated and the optimum locations for the number of facilities available were suggested (Godlund 1961; Abler et al. 1971). In the USA, Pyle (1979) discussed the geographical content of areawide planning and subsequent programmes. In Britain, geographical aspects of health services distributions have become recognized most clearly since reorganization of the NHS (Department of Health and Social Security 1976a).

A general aim of planning may be assumed to be to achieve optimum access and utilization of a given service for an intended population, although planning may, of course, have numerous other service-related aims such as redistribution and reallocation to favour certain groups or areas. In a complementary fashion, social planning may be seen as rational behaviour among groups, the process through which societal goals are pursued. Health care planning is a subset of this social give and take. In principle, social planning requires the formulation of goals and objectives and the identification of alternative planning strategies for each planning subset. Unfortunately, this has frequently not been the case. Few countries have explicit social aims; most are ad hoc, temporary and expedient. At times, health care in Britain, for example, seems to have been planned largely in isolation from, and even in contradiction to, other aspects of public policy (Eyles et al. 1982).

Therefore, is there any point in attempting to seek order in health care planning? The practical answer is that an attempt to do so must be made because of the importance of health care to people's lives and because of its pre-eminence as a public service, especially in cost terms. Locational aspects of general public policy are well discussed elsewhere (Massam 1975, 1980; Cox 1979). However, how can the main foci of health care planning be viewed? Scale can provide a useful analytical framework.

At the international level, health care planning has mainly been very general. Sometimes, all that has been achieved is a collation of data on personnel and facilities, on utilization and on medical care supply:population ratios, emphasizing the disparities amongst nations (Kohn and White 1976; Coates et al. 1977; Cox 1979). At other times, however, it has achieved specific health-oriented goals, perhaps really part of 'health' planning. Campaigns to eradicate smallpox, to combat cholera, malaria and other infectious diseases provide excellent examples and some are success stories in international co-operation in health control (Howe 1977; Learmonth 1978). At this scale, the World Health Organization is a major co-ordinating body.

Some commentators are worried, however, that the WHO's good intentions are often cloaked in eloquent but high-flown verbiage (Barley 1982). Perhaps this has lessened the impact of a recent campaign, otherwise a good example of setting goals in health care planning and suggesting some means of achieving them. *Health for all by the year 2000* is the global strategy of the WHO member states, a collective goal, but one based on national policies and strategies. The 1977 World Health Assembly had decided that 'the main social target of governments and the WHO should be the attainment by all people of the world of a level of health that will permit them to lead a socially and economically productive life' (World Health Organization 1981, p. 11). This was followed in 1978 by an international conference on primary health care in Alma-Ata, USSR – a practical outcome of the 'health for all' strategy. Then, in 1979, the strategy and the Alma-Ata Declaration and Report were endorsed, the member states being invited to act individually in formulating national strategies, and collectively for regional and global strategies.

As discussed in Chapter 3, primary health care constitutes the foundation upon which the health care hierarchy is built. International declarations such as that at Alma-Ata have helped to promote it as a concept. It is, indeed, seen as the key to attaining the goal of health for all. Good primary care is viewed as including a network of social, educative and preventive measures (nutritional, clean water, mother and child care, and immunization, to name but a few) and to involve all sectors of the nation and community. The Alma-Ata declaration calls for a 'deprofessionalization' in so far as all types of health care workers and auxiliaries, including traditional practitioners (see Chapter 2), can be included. It calls for *essential* health care, based on appropriate technology and socially acceptable methods.

Do these grand designs, eminently reasonable though they be, have practical planning application (which is arguably the measure of good planning)? In the translation of high-flown language into everyday speech, the WHO has had a stimulating effect because it has pointed out forcibly that in some parts of the world people are much less healthy than others. This disparity is unjust and also dangerous, because health, peace and development depend worldwide on each other. If things are to be put right by the end of the century, then nations must work together and should devote more resources to caring for the ill, eradicating disease, housing and feeding people properly, and generally educating them in how to live healthy lives (Dukes 1978). If this is to be achieved, international co-operation is essential, so this scale of planning, albeit sometimes of a rather sermonizing nature, is invaluable.

At the national scale, countries build health care infrastructures and employ manpower, and they educate, feed and clothe their people. Is this, therefore, the most appropriate scale for health care planning? The answer can only be: for some purposes. Resource allocation in welfare states may take place at a national level but, in many countries, resources are not centrally generated or allocated. It is sometimes difficult in such instances to envisage a process of 'planning' since much of the health care system may be provided by elements not included in a government-sponsored design.

There are examples of geographers contributing to the locational strategies for health services at a national level although this has been more commonly achieved at a regional or urban level. Horner and Taylor (1979), for example, provide an evaluation of various proposals for reorganizing the Irish hospital system. At various times since the Fitzgerald Report of 1968, 12, 18 and 22 sites had been proposed and, using a grid-square process to achieve travel minimization for potential users, the suggested locations were compared with optimal locations (Tornqvist et al. 1971; Robertson 1978). Levels of adjustment of sites and travel times and distances were compared for various solutions, as illustrated diagramatically for the 12 centre solutions in Figure 8.1.

This sort of procedure provides the decision-maker with the basic data to identify and evaluate alternative locational strategies. He must still, however, decide what weight to allow to these locational data in relation to any other considerations. Equality

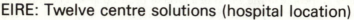

EIRE: Twelve centre solutions (hospital location)

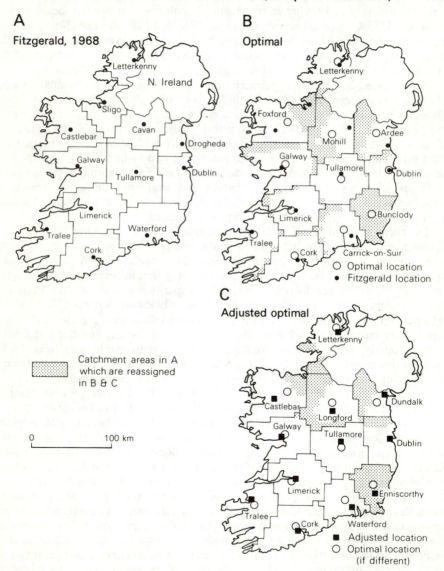

Figure 8.1   Hospital locational strategies in Eire

Source: after Horner and Taylor (1979)

of travel (or travel minimization) may save on personal costs but may not be justified if there are some very needy areas where resources should be concentrated. In addition, rarely will the decision-maker be faced with deciding freely upon the locations for new facilities as he will often be tied by existing sites or by more general interfacility or even interservice linkages (White 1979). At a regional scale, Haynes and Bentham (1979) provide a good example of how locational features can be weighed in a model along with others to provide a number of alternative locational strategies for community hospitals. At a subregional scale, Bennett (1981) and Curtis (1982) provide examples of distance-minimizing solutions for access to primary health care. Local-scale planning is considered later in the chapter.

Planning constraints and strategies

Within health care systems which are mainly government-provided or at least ordered to some state or national plans, E.G.Knox (1979) envisages a circular planning process operating in a socio-economic environment containing financial, manpower, legal/ethical and other constraints. A number of stages succeed each other and, as each stage raises fresh questions, so a cycle develops (Figure 8.2). Of critical importance is the situation analysis in which the correct problems are identified and tackled. If important issues are by-passed or ignored at this early stage, only partial success at best could be achieved by even the most efficient planning system. It is here that society's (value-laden and culturally determined) goals and objectives exert their influence most potently. If, for example, an underlying goal of the nation is to provide health care for all, this will imply the creation of a very different health care system from that which would develop in a nation where the aim was to provide health care according to financial means or for certain groups only.

Health care strategies may be evaluated in very different ways (at a national or more local scale) and this links with the following section. 'Outputs' of a service can be measured in terms of cost-effectiveness, efficacy or efficiency but they can also be related to equity and equality, if it is a goal to achieve 'targets' on either of these objectives. It must be kept in mind, however, that whereas the planner *forecasts* potential outcomes of alternative policies and may make comments on their potential outcomes or evaluation in advance (E.G. Knox 1979), it is *society* that determines which outcome it feels is more desirable.

Local strategies

The *local* scale is often the most appropriate for day-to-day planning. It is at this scale that questions of accessibility and utilization become practical. The service has to be 'delivered' at this level: the 'consumer' has to be able to have access to the service both physically and socially. It is important here to distinguish between two key concepts. *Accessibility* is by and large a measure associated with places and locations whilst *mobility* relates to the characteristics of individuals or households (Moseley 1979). Both tend to vary according to the geographical characteristics and the socio-economic differentiation of an area, such as levels of car ownership (Thomas 1976). A health facility's location in a district provides another (origin and destination) node in an already complex interaction system, and lack of attention to local circumstances of accessibility and mobility can aggravate what could already be an unsatisfactory state of affairs in terms of travel distance, cost, and time.

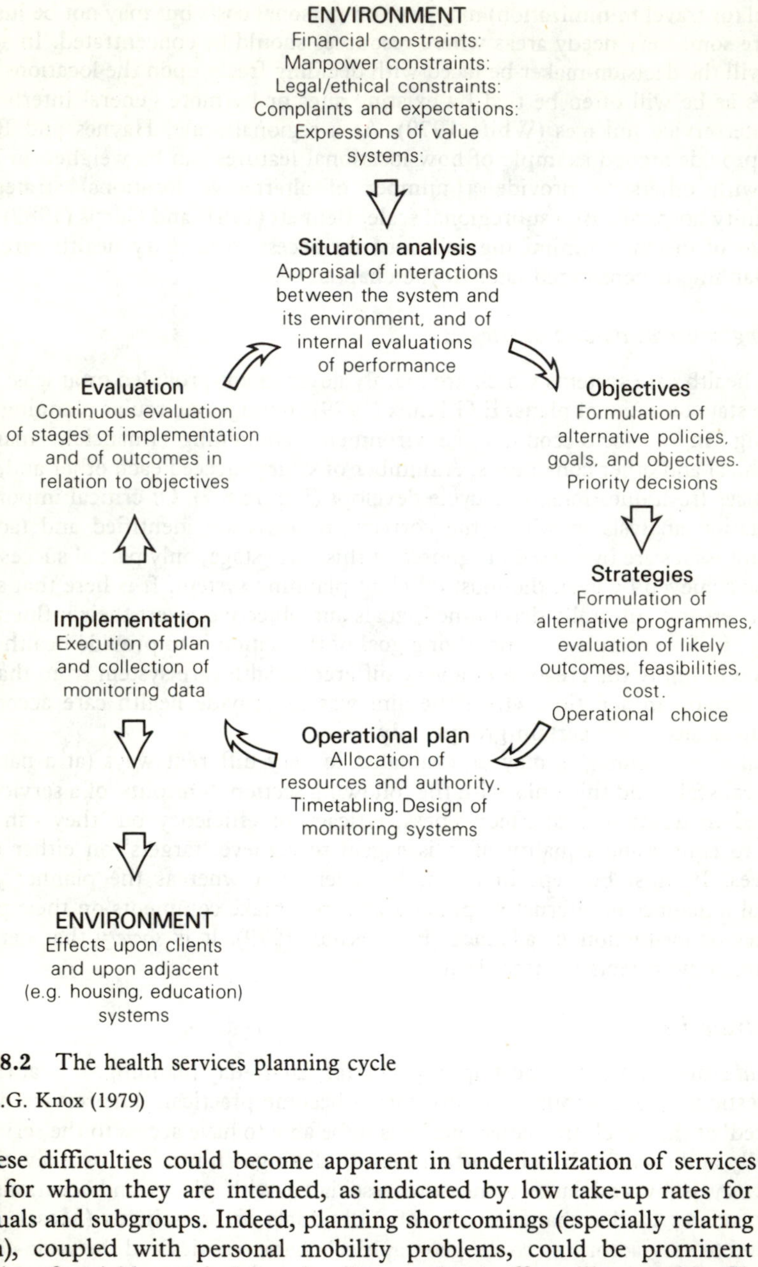

ENVIRONMENT
Financial constraints:
Manpower constraints:
Legal/ethical constraints:
Complaints and expectations:
Expressions of value
systems:

Situation analysis
Appraisal of interactions
between the system and
its environment, and of
internal evaluations
of performance

Evaluation
Continuous evaluation
of stages of implementation
and of outcomes in
relation to objectives

Objectives
Formulation of
alternative policies,
goals, and objectives.
Priority decisions

Strategies
Formulation of
alternative programmes,
evaluation of likely
outcomes, feasibilities,
cost.
Operational choice

Implementation
Execution of plan
and collection of
monitoring data

Operational plan
Allocation of
resources and authority.
Timetabling. Design of
monitoring systems

ENVIRONMENT
Effects upon clients
and upon adjacent
(e.g. housing, education)
systems

Figure 8.2   The health services planning cycle

Source: E.G. Knox (1979)

These difficulties could become apparent in underutilization of services by the groups for whom they are intended, as indicated by low take-up rates for certain individuals and subgroups. Indeed, planning shortcomings (especially relating to poor location), coupled with personal mobility problems, could be prominent in the composite of variables underlying the distance decay effects illustrated in Chapter 6. Trips and visits which are forgone or delayed can later result in extra costs to individuals and to society. Therefore, the significance of the local level of planning is to be emphasized. 'At a national level, it is possible to reflect on broad issues and produce general conclusions (of real value) whilst, at the local level, planning has to take place against a background of actual buildings, staff in post, geographical distributions and, often, long histories of a particular kind of provision' (Grime and Whitelegg 1982, p. 202). Because these are very complex issues which vary locally, national level 'broad bush' planning cannot concern itself with them. As discussed subsequently, certain health authorities have attempted to reconcile these two levels, which is the essence of good and effective planning.

What is planned for health care: efficiency, equality or equity?

Equality means that every person receives the same treatment or share of resources regardless of who or where he or she may be. Arithmetic equality suggests that everyone receives exactly the same, whilst proportional equality can mean that people receive things in proportion to preset norms: equal treatment in the same circumstances where these can vary among groups and individuals. Equity implies *fairness* in distribution, suggesting recourse to principles of natural justice, in which people or groups may be given differential treatment if it is thought just (Smith 1977, 1979).

In many societies and almost invariably within the Western world, equity and equality are equated with one another (see Chapter 3). There is a certain logic in this although it can also be argued that equality of outcome is more important than equality of inputs (Coates et al. 1977).

Geographers have fastened onto the idea of social justice as a way of evaluating the distribution of phenomena such as income, wealth, education, shelter and health care in spatial terms, and 'territorial social justice' was discussed by Harvey (1973). However, it has been surprisingly poorly researched for such an important topic (Pirie 1983). Pinch (1979), Smith (1974, 1977, 1979) and a few others have provided empirical examples of territorial justice, the anlysis of who gets what, where and how. Perhaps the notion of territorial justice should now be more convincingly extended into a fully fledged concept of spatial justice (Pirie 1983). However, there are problems in using solely an areal framework for the evaluation of social justice, and the shortcomings of area-based policies, such as the implicit assumption that problems will be concentrated in identifiable locations and hence be amenable to blanket area-specific remedies, are now recognized (Hamnett 1979). In the sense of utilizing a territorial reference point for justice, this is quite appropriate, as Pirie explains, for comparative regional research and for forming some policies. However, within territories, it is often necessary to know about the 'justness' of distributions, and 'best-location' solutions such as those discussed in earlier chapters are often seeking answers in terms of maximizing accessibility or some other criterion, rather than in terms of fairness.

From a managerial viewpoint, *efficiency* may have little to do with equity, equality or social and spatial justice. It can often mean achieving the best buildings per unit of currency expended, or the minimum waste of physician or nurse time. The 'welfare approach' in human geography realizes this and has attempted to look at the results in terms of inequities in the distribution of goods for clients (Smith 1977) although the influence this has had on health care planning to date in any nation is not clear. In statistical terms, certain indices (such as gini coefficients) are available which can measure the difference between an equal distribution (say, of health care investment) and an actual distribution identified by a Lorenz curve. An example of a gini coefficient was given in Chapter 5 in the form of the coefficient of localization for GP services. Such information can provide health care planners with at least a quantifiable indication of regional maldistribution of resources, and policies can then be put in motion to adjust this if it is felt to be inequitable or counterproductive to strategic aims. A number of examples can be cited of attempts to redress various types of imbalances in distribution of health care resources (see Chapter 4). However, most have relied on relatively simple population:practitioner ratios to monitor system performance in, for instance, general practice (Butler and Knight 1974; Salmond 1974; Bryant 1981) and dental care (Ashford 1978; Scarrott 1978; Carmichael 1983).

An example of a largely managerially oriented system (as opposed to one which is seeking territorial justice) is to be seen in the planning of health services by rather strict numerical norms in Hong Kong. The Hong Kong Outline Plan (now called the Hong Kong Planning Standards and Guidelines) provides detailed information on the nature and levels of services that will be provided in new residential developments (Phillips 1981b). In particular, norms are laid down for medical and social service provision in public housing estates which will provide homes for over 2 million persons in a series of new towns now under construction (the main current thrust of public housing in the territory). Some examples of the rather mechanistic standards are given in Table 8.1. For example, planning standards previously specified 1 GP per 6,000 residents, but were recently amended to 1 GP prt 7,500–10,000 residents. Little or no account is taken of factors such as the age structure or socio-economic status of residents although it is recognized that these can affect considerably their use of medical services. In addition, traditional Chinese medical practitioners are not explicitly included in the planning of medical services because the Hong Kong Government provides only Western medical facilities. Therefore, the extent of utilization of traditional medicines is not explicitly anticipated although it is realised that a certain usage will occur. This combined with the lack of behavioural data on utilization can lead to relative undersupply or (possibly) even oversupply in terms of total medical services available and required (Phillips 1983a, 1983b). Hong Kong provides an interesting example of planning where services are often being provided in totally newly built environments such as new towns or new public housing estates. It also provides a fairly extreme example of 'planning by numbers' in a non-socialist state, necessitated by the impressive scale and speed of its urbanization. However, overall, it seems that this system is attempting to provide equality of distribution (perhaps to ease physical and financial planning), rather than to achieve equity in health services provision.

One major problem in determining a just geographical distribution of health care is that there is frequently little clear idea of what the medical needs are in different populations; neither is there a fixed norm of how many patients a doctor can deal with (Butler 1976; Royal College of General Practitioners 1972). It is often expected that young families and the elderly will need more care but many needs may not be expressed as demands because of locational, social or financial 'inaccessibility' of services or low levels of perceived need (see Chapter 6). Therefore, demand (or even morbidity) may be a very poor indicator of medical need and can accidentally lead to the planned underprovision of medical manpower and facilities.

Resource Allocation Working Party (RAWP)

In Britain, an elaborate planning excercise referred to in Chapter 2 was initiated in May 1975 in the recognition that there were very conspicuous differences in levels of

Table 8.1 *Selected Hong Kong Outline Plan standards for social welfare and community facilities (Government and privately provided facilities: standards planned in new developments)*

Facilities	HKOP Standard*
Medical facilities	
Hospital beds	target 5.5 beds/1,000 persons
Clinics	1 doctor/7,500–10,000 persons
Health centres/Polyclinics	Flexible
Educational facilities	*Bisessionally-used classrooms*
Kindergarten (not statutory) } serving area of	1 classroom (90 places)/2,700 persons
Primary } 0.4 km radius	1 classroom (90 places)/850 persons
Secondary } 3.2 km radius	1 classroom (unisessional, 30 places)/520 persons
Special schools	
Higher Education	not specified
Community facilities	
Child care centres/Kindergarten. No set standard	Social Welfare Dept. suggests 100 places/20,000 persons
Youth centres	1 centre/20,000 persons
Children's centres	1 centre/20,000 persons
Social centres for the elderly	1 centre/30,000 persons
Estate community hall	1 hall/20,000–50,000 persons
Estate community centre	1 centre/50,000 persons
Community centre	1 centre/100,000 persons

* Figures are for places per total population

Source: after Phillips (1981b) — norms updated to 1982 guidelines services among regions on a number of indicators such as list sizes, expenditure and morbidity (Howe 1963; Coates and Rawstron 1971; Rickard 1976). RAWP had the very difficult task of reviewing the distribution of capital and revenue among regional health authorities (RHAs) with a view to establishing a method of achieving a pattern of distribution responsive objectively, equitably and efficiently to relative needs (Department of Health and Social Security 1976a). Such terms are themselves not unequivocal, of course, but the aim of creating equality of access to NHS health care for people at equal risk was laudable if, perhaps, unattainable.

RAWP attempted to place RHAs relative to each other in terms of 'revenue targets' determined for each: some were 'overfunded', being above their targets; others underfunded. The revenue targets are, of course, highly debatable but were achieved by a complex formula based on variables such as population, age, mortality and fertility. The revenue available nationally was distributed in proportion to each authority's combined weighted population, and this could be determined for each sector of the health service. The relative position of authorities can be illustrated as in Figure 8.3, and it was RAWP's intention to allocate resources regionally so that, ultimately, all authorities would lie along the horizontal line. The effects of this national scale allocation exercise have been to divert resources away from the South-East (London-Thames health authorities in particular) and to the North and the West. At an intraregional scale, the major impact has been to divert funds away from inner-city areas (Eyles et al. 1982; Eyles and Woods 1983).

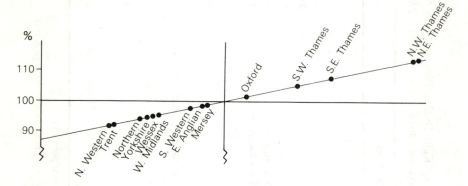

Figure 8.3 RAWP: relative positions of Regional Health Authorities in relation to revenue targets (RHAs above the line may be regarded as 'overfunded')

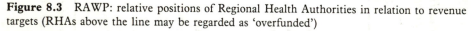

Source: Department of Health and Social Security (1976a)

RAWP has been criticized on philosophical, technical and practical grounds. The use of apparently sophisticated 'scientific' analysis can give the appearance of pseudo-rationality and fairness. Because of its aim of providing greater equality amongst authorities, RAWP achieved redistribution of resources from supposedly overfunded to underfunded authorities. It divided a finite 'cake' differently, but it did not increase the size of the 'cake'. It can therefore be argued that the exercise limited the growth of real (State) expenditure on the NHS. As Eyles et al. (1982) point out, this illustrates the contradictory nature of bureaucracy and planning in complex mixed (free market/welfare state) societies. The increasing expenditure and intervention of the

State can appear to threaten the existence of the private sector (in health, education, transport, housing, and so on) and thereby threaten capitalism itself. As a result it becomes necessary to retrench on public investments and release funds for private expenditure which, almost inevitably, will be distributed unevenly.

Technically, the criteria used by RAWP in its population weightings can be criticized. It is evident that utilization and standardized mortality ratio weightings are only imperfect indicators of need, particularly since they give little consideration to social deprivation, especially important in some large cities (Avery-Jones 1976). It is unclear, for example, whether a mortality rate of 10 percent above the average imposes a 10 percent greater burden on health services and, indeed, whether this therefore requires a directly proportional increase in resources. However, Eyles et al. (1982) suggest the greatest technical deficiency of RAWP is that it says little about the expenditure patterns among service *sectors* within health districts. Within inner-city districts, a key issue is how to transfer funds to community and primary services, away from 'acute' care and large hospitals, which have considerable inertia because of vested interests and existing infrastructure. In one notoriously disadvantaged London Borough, Tower Hamlets, a required reduction in hospital bed numbers was to be achieved by closing beds in small hospitals serving *local* needs while at the same time permitting the concentration of acute services in a larger unit, in the London Hospital (Yudkin 1978).

In practical terms, the reception which RAWP got depended on the position of individual authorities regarding future financing. Those with most to lose were the most critical, particularly the inner London boroughs containing the major teaching hospitals. Suburban health authorities would generally give guarded approval. This type of planning, along with that of the London Health Planning Consortium, tends to isolate health care, and health services in particular, from more fundamental questions facing inner-city authorities. Eyles et al. (1982) argue that these should ask what the *health care needs* of populations are and *what types and combinations of services will best meet them*. Instead, it may be argued that bureaucracies become self-sustaining, their own needs rather than those of the population dominating planning (McKinlay 1972; Eyles et al. 1982). Acute hospitals and overfinanced authorities, for example, will be unwilling to give up their resources or to effect the supposedly needed redistribution. Co-ordination also becomes a problem and, in Britain for example, there is still an overlap and poor liaison between NHS primary care and local authority personal social services. Although there are supposed to be channels of liaison and means of jointly funding projects of common interest (e.g. for elderly or handicapped persons), a lack of will and bureaucratic delay almost seem to formalize the lack of co-ordination. As a result of this general orientation, it has been remarked that the procedures adopted under the NHS have at times contradicted other aspects of government policy, especially with regard to the inner-city, into which other government departments have been putting resources whereas the NHS has been withdrawing resources from them. 'The planning of health care provision (including spatial resource allocation) thus proceeds largely in isolation from other aspects of public policy' (Eyles et al. 1982, p. 251). The aims of the NHS and of reorganization do not yet seem to have overcome this.

Two other important issues relating to equity and equality of health care have implications for access and utilization: first, the problem of measurement of 'quality' in health care and, secondly, the provision cost explosion in recent years.

Quality of health care

The measurement of 'quality' is the chimera of health care. Does 'good-quality' care cure more illness or do high levels of illness to be cured indicate poor preventive care? Do 'good' GPs send more or fewer patients to hospital? Is a young, technically well-qualified doctor 'better' than an older, more experienced physician? Is recourse to paramedical personnel 'good' because it releases the physician for more serious duties, or does it indicate a type of laziness? Many questions such as these can never be resolved and, indeed, although 'bad' medicine is costly to individuals and to a country, this is hardly measurable even on quasi-objective indices such as life expectancy or morbidity (Honigsbaum 1972).

However, resources are often allocated in terms of these indices. Reference was made earlier to the British method of 'allowing' GPs to work in certain types of areas based upon the criterion of list size (patients per doctor). In areas with small numbers of patients per doctor, new practices may not be permitted (Butler and Knight 1974, 1976; Phillips 1981a). This makes no allowance for 'quality', which may be illustrated by the case of the Thornhill district of inner London, near King's Cross. On 'objective' criteria this area is underprivileged, suffering from multiple deprivation with above-average levels of infant mortality, malnutrition and disease in a population with a high dependency ratio of young and old. There are also above-average numbers of vagrants, alcoholics, handicapped persons and ethnic minorities, all of whom might be expected to have special health care needs (Thornhill Neighbourhood Project 1978).

In numerical terms, Thornhill appears to be well provided with GPs and physical accessibility to primary care is high: 21 GPs work in 13 surgeries in the 1.5 square mile neighbourhood. Therefore with various population adjustments, the average list size is fewer than 1,800 per GP, and the neighbourhood is therefore closed to all new practitioners on the grounds of overdoctoring (see Chapter 4). However, if 'quality' of care is taken into account, Thornhill is medically deprived. The age-structure of the doctors is extremely skewed. Over one-third of all the doctors were over 70 years of age at the time of the study; two practitioners, whose training must have taken place some years before the discovery of penicillin, were over 80! Knox (1979a) questions whether, in spite of the value of experience, practitioners of this age can effectively tackle a full work-load in an area with intense and specialized needs. Many of the older GPs do not attempt to do so and deliberately restrict the size of both their lists and catchment areas. In addition, many doctors work in older, poorly equipped surgery premises, often finding it difficult to employ a nurse or a receptionist. Therefore, primary medical care in Thornhill is rarely being provided in group practices with ancillary help from a primary care team – the officially encouraged and desirable method of provision (Royal College of General Practitioners 1972, 1979; Phillips 1981a).

Thus, an area which seems to be well provided with medical manpower on apparently objective planning criteria turns out to possess an unco-ordinated collection of GPs, working from undercapitalized surgeries. This is not a unique case as Knox (1979a) illustrates for several Scottish cities. In Glasgow's East End, apparently high levels of physical access to GPs are devalued by poor-quality care in a harsh and deprived inner-city environment. There are high levels of unemployment and an ageing population, together with numerous vagrants and other vulnerable groups with special medical needs. However, primary medical provision here, too, is of relatively poor quality: Knox found that some three-quarters of GPs were over 45 years of age and 17 percent over 65. Over 40 percent were working in solo practices, many in lock-up surgeries open only during consulting hours and in grubby premises with little diagnostic equipment. As a result, many patients seek primary care from the local hospitals, not strictly permitted under NHS rules and undesirable because of lack of continuity and because it constitutes a shifting of the problem from one NHS sector to another rather than a real solution.

These studies indicate the problems of assessing quality in care without also looking at a range of social indicators and measures of need. This provides a warning that countries with centralized or State-run health services may ostensibly be well provided but the quality may be in doubt. The major State-run health services of socialist countries have particular problems related to the achievement of 'high-quality' care whilst relying on work norms and allocation by number (Ryan 1978). This is one of the intractable problems of health care planning and it is not even overcome in free-enterprise systems, as 'poor-quality' care may merely be avoided by some patients who can afford the 'best' care available.

Health care provision: a cost explosion?

Many countries have noticed enormous increases in the costs of providing health care in recent years. Virtually all overviews of health care delivery refer to this phenomenon, which has been dubbed a 'cost explosion' (in Germany, 'the Kostenexplosion'). Some of these increases have no doubt been caused in part by defective planning: for example, the provision in some European countries of giant hospitals when there are no public funds to staff and run them! Other increases have been attributable to inaccurate projections of population growth or to misconceived treatment norms.

Attitudes concerning health care have begun to change with new economic realities. A generation ago it was claimed by many that it was a fundamental right of all people to have access to the best possible forms of health care. This nonchalant attitude to costs continued through the growth and expansion of hospital services in the 1960s but, by the late 1970s, the mood was very different and, in many countries, politicians and planners were asking how much is enough. Has the cost increase in health care resulted in improvements in health status anything like commensurate with the extra expenditure (Vuori 1982a)?

The consistency of the cost increase trends internationally is remarkable, and costs have usually increased ahead of inflation. In Japan, the costs of medical care in 1960 amounted to a total of 410 billion Yen, rising by four times by 1970, and multiplying again by 1979 to 10,951 billion Yen. This rate of growth was matched by an increase in the percentage of GNP absorbed by medical care. The picture is the same elsewhere, as illustrated in Chapter 2 (Table 2.1). As long as Japan was able to achieve its phenomenal economic growth rate of 9 percent per annum in the 1960s and early 1970s, such costs could be absorbed but, with the international economic slow-down of the late 1970s and early 1980s, substantial increases in contributions by employers and employees have been needed (Chester and Ichien 1983). In New Zealand, the same question is asked: how to cope with a fourfold cost increase between 1969 and 1978 in view of the adverse conditions affecting the national economy (Heenan 1980). Other international examples of consistent increases in the proportion of GNP or GDP absorbed by social and medical services were given in Chapter 2 (see Tables 2.1 and 2.2).

The same five causes have been almost universally identified as underlying the rise in health care expenditure (Maxwell 1980):

1 *Changing needs,* particularly of ageing populations, and associated increases in the prevalence of chronic disease.
2 *Technological change,* since medical advances offer the possibility of improved care but rarely save money.
3 *Rising expectations* on the part of health care professionals and the general public, who expect better standards of acute care and also of care for groups such as physically and mentally handicapped persons, often neglected in the past.
4 *Rising personnel costs.* Health care is labour-extensive; rarely is human labour replaced by capital equipment. Indeed, technical advances (such as computer tomography) usually require additional trained staff.
5 *The increasing dependence on public financing* can lead to fewer personal and professional scruples about costs and charges.

Virtually all of these causes will continue in the future although the last has perhaps had its maximum impact because all governments have now become concerned about the cost of health care and seek means to limit public expenditure. Britain, Canada, France, Sweden and the USA, amongst developed nations, have all attempted to limit health care expenditure, with relatively little success. Major questions for governments must now be how to obtain the maximum value for money in expenditure on health and how to choose rationally amongst health programmes and other priorities. This is a keen concern of politicians although not always recognized by health care professionals. So far, two foci of concern have emerged (Vuori 1982a):

1 The nature of the balance of different components of health care systems.
2 What should the proper balance be between health services and other services, such as education, housing and nutrition, which contribute to health promotion?

The first focus has led authorities, particularly in many developing countries, to question the wisdom of high rates of capital expenditure on hospitals and high-technology developments. Better value for money in terms of cases treated has often been achieved with more primary health care and with self-help and community initiatives. In virtually all countries, hospitals account at present for a major proportion of health expenditure, ranging from 71 percent in Sweden and 60 percent in Britain, to some 40 percent in France and West Germany (where costs are low owing to the nature of outpatient care accounting in these two countries). They have often been rising ahead of other health care costs (Pyle 1979). In many developing countries, prestige hospital developments can dwarf the national health budget and serve relatively few people (see Chapter 2). Perhaps the closing years of this century will see a worldwide change in this hospital orientation.

The second focus has investigated the ways in which cheaper medical care can be a substitute for more expensive services. In particular, preventive medicine and health education are very valuable as they may reduce the need for future health services (Department of Health and Social Security 1976b; Sutherland 1979; Learmonth 1982). Health care is increasingly recognized to be part of the wider fabric of social provision, including, for instance, day centres for elderly people, hostels, and residential homes for mentally and physically disabled persons. Therefore, strengthening social service provision and alerting the health professions to the possibilities of care in these sectors as viable alternatives for medical intervention can go some way to reducing the future cost explosion.

Health care planning for special groups in society and some planning initiatives

There is increasing recognition of the special health care needs of subgroups in the population. Apart from ethnic minority groups, there are probably five priority subgroups for whom health care planning has special concern: elderly persons; disabled and non-mobile persons; mentally ill persons; mentally handicapped persons; and children. The last named group has been the subject of numerous studies in relation to health services and needs, such as the Court Report (1976) in Britain, and the Child in the City Programme in Toronto (Andrews 1982). Space does not permit a full discussion of the special health care needs of all these groups, nor their implications for access and utilization. However, an example is provided here in planning health services for elderly persons, and this also introduces certain local planning initiatives which have been developed.

Elderly persons

The proportion of elderly persons is rising in the population of almost every country although north-west Europe is the first region in which most countries have more than 12 percent of their population over 65 years of age. Research has focused on the regional distribution of retired persons (Law and Warnes 1976; Golant 1979; Wiseman 1979; Warnes 1981, 1982), their housing (Barnard 1982; Heumann and Boldy 1982), their living environments (Golant 1979), and their health and morale (Lawton 1970; Herbert and Peace 1980). Old age frequently involves increasing disability, loss of mobility and increasing dependency, and therefore the planning of services for elderly persons is important and is made particularly pressing by increasing numbers.

In addition to the increasing number and proportion of persons aged 65 and over, there has been an increase in the proportion of elderly persons living alone. Fewer family members are available in Western societies in particular to provide 24-hour support at home, and in many cases those surviving to old age are widowed or unmarried women. In the USA, two-thirds of all women aged 65 and over now live alone or with someone other than a spouse (Treas 1977). In Chapter 6, age was clearly indicated as a variable influencing the utilization of medical services, and the old *and* alone may produce even higher rates. It is prudent, therefore, for medical planners to take very careful account of changing demographic patterns as well as economic and technical matters when future facilities are designed and located. For example, health centres in neighbourhoods of average, elderly and juvenile age structure would have very different health care needs and utilization patterns. These will influence the types of professional training and services required, and this should be programmed at an early planning stage.

Health services planners frequently implement 'situation analyses' of existing demographic structures, with future projections made on the basis of known demographic trends. Subsequently, it will be realized that certain problems can be met only in terms of national priorities; others will require local solutions. At a policy level, the changing proportion of elderly infirm persons, for example, may dictate that more residential places and greater domiciliary support are provided to reduce the demand on institutional support. Medical institutionalization is seen increasingly to be inappropriate for a large number of the elderly 'sick'; many studies have identified institutionalized elderly persons who could live semi-independently were appropriate long-term housing provided in the community (Heumann and Boldy 1982). However, increasing the availability of sheltered housing may require new building or adaptation of existing homes. Resources may only be available for such initiatives if patients who do not need hospital are discharged and the costs of those who must stay in hospital are reduced to ensure their facilities are not serviced more than a reasonable standard of care demands. Such decisions may be best made at a national level. In addition, an effective system of care in the community demands that admission to medical or other institutions only be for specific circumstances rather than as a general expectation, and that domiciliary support and preventive surveillance can forestall critical episodes.

The timely formulation of strategies to provide for appropriate levels of community care and sheltered accommodation is essential if societies are not to uncaringly abandon their old folk to their own resources. Studies in Australia (Lefroy and Page 1972), the United Kingdom (Shaw 1975; Herbert and Peace 1980) and the USA (Golant 1979) have shown that elderly populations living in their own homes may contain high levels of functional and physical disabilities, social isolation and service dependency (with low actual levels of support). In addition, E.G. Knox (1979) suggests that many people in institutional care are not in the situations best suited to their needs. Many in hospitals would be better in residential homes and vice versa. Some in hospitals could be discharged into the community were more attention paid to supporting services, whilst others living at home might realistically be better off in residential care.

Therefore, both situation analyses and strategic planning are required to prevent problems developing. At a national scale, resource allocation must reflect the changing proportions of elderly persons. At a subregional or urban scale, the planning must reflect matters such as transport and service availability and the mix of housing (Schmitt 1979; Birdsall 1979; Barnard 1982). This must especially take heed of particularly needy or disadvantaged elderly persons such as the poor or those in rural locations (McKelvey 1979; Mercer 1979; Phillips and Williams 1984). At a micro-level, planning must make housing and local environments appropriate for elderly persons whose mobility and length of journey may be restricted, taking note of local features and obstacles such as hills, steps and housing design (Millas 1980; Heumann and Boldy 1982). It is in service location for specific populations that some of the techniques outlined in Chapter 5 can be effectively utilized, especially at an intraregional level.

Ideal planning resolutions for subgroups are rarely achieved, however, because of national and local hindrances to commonsense solutions. The administrative responsibilities for supportive services are often fragmented and their liaison poor, so that, in the absence of sufficient increases in resources, better co-ordination is essential. This may be achieved by a spatial planning framework that enables administrative collaboration at a small spatial scale.

Locality planning

In Devon, England, an experiment in locality planning has been implemented by the Exeter Health Authority since early 1981 with the aims of improving liaison and heightening interprofession awareness of problems and the potential for sharing resources. The 300,000 population health district was divided into localities ranging from 10,000 to 30,000 persons, and 'local consortia' of all those involved in planning and delivery of health and related services began to meet together informally to discuss common problems and to suggest remedies (Figure 8.4). The experiment was unusual in that all those involved in the broad sphere of 'health services' were included: the NHS, GPs, community nurses, social services, housing, pharmacists, home helps and some voluntary organizations. The 16 locality meetings were co-ordinated but not dominated by the Exeter Health Authority, and explicit in the idea was that participants would be 'grass roots' workers involved in day-to-day practical care. This worked well in the initial meetings and it has now been officially incorporated as a practical approach for future planning by the Exeter Health Authority and the local social services department.

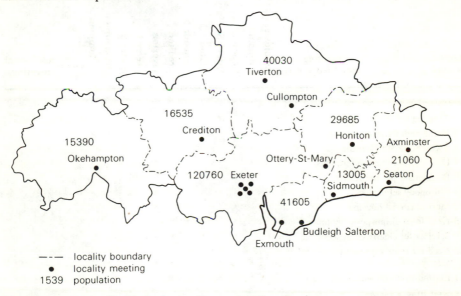

Figure 8.4   Exeter Health Authority, Devon: locality planning

The Exeter experiment seemed to work because field workers at the local level were being asked directly for their views on problems, resources and manpower. It became apparent that a series of common problems existed, often relating to provision for elderly persons in this popular retirement area (Law and Warnes 1976). Once this became apparent, the burden could be shared and the 'consumers' more rationally treated by their health care professionals who more readily accepted transfers owing to the better liaison established at the locality meetings (Phillips and Court 1982).

This method could be adopted and extended in many countries where health and social care is provided by a variety of agencies. Since sharing and sensible allocation of resources is often *the* major problem (and often underlies the cost explosion), increased interagency liaison is now increasingly urgent. The Exeter methodology was to identify common themes which were raised at meetings and then to promote these as a basis for 'grass roots' planning. An example of the type of tabulation which can be constructed is given in Table 8.2. The notion was to plan in the knowledge of locally available resources and locally perceived problems rather than to impose plans based upon the view from the upper hierarchical levels of the health and social services bureaucracies. The involvement of those providing care locally in the future planning of services was surprisingly novel and, in the Exeter case, the consumer was also represented by a parallel survey of rural health facilities by the local Community Health Council, the results from which were provided to the planners (Exeter Community Health Council 1983). This commonsense approach to local planning should optimize utilization, not necessarily by maximizing it, but by ensuring that

Table 8.2 *Locality planning: examples of topics which can be raised at meetings*

Topic Raised	Locality meeting (see Figure 8.4)	Axminster	Sidmouth	Honiton	Ottery St Mary	Seaton	Exmouth	Crediton	Tiverton	Exeter 1	Exeter 2	Okehampton	Cullompton	→ additional meetings
Elderly														
1 Training for staff		*	*		*		*		*					
2 Home help duties		*					*					*	*	
3 Twilight nursing scheme					*					*				
4 Day centres linked to hospitals				*	*									
5 Domiciliary physiotherapy				*		*								
6 Chiropody services				*	*	*		*	*			*	*	
7 More sheltered housing			*				*	*					*	
8 Need for multipurpose day units		*	*				*	*	*				*	
9 Services for elderly confused			*			*		*	*					
10 Need for team approach to care for elderly				*		*	*		*	*	*	*		
11 Expand meals-on-wheels service				*		*			*		*			
Mental illness sector														
1 Need for more community psychiatric nurses						*		*	*		*		*	
2 Joint user multipurpose day units					*		*			*		*	*	
3 Day/residential care for adults		*	*				*							
4 Hostel for young mentally ill					*									
5 Play-group					*									
6 Improve local hospital services		*					*							
7 Crisis intervention team						*						*		

Note: Topics raised at meetings (*) are purely illustrative but show the range from local to districtwide planning issues health services are used by, and available to, those requiring them. Accessibility can be improved by reducing waiting lists and by identifying locally underprovided settings which can be given priority in subsequent facility provision. Locality planning is of considerable interest to health care geographers as it is inherently spatial, the designation of natural localities being a crucial stage of the execution, and has the essential features of practicality and flexibility.

In addition, locality planning may help to overcome a lack of conformity between national health care planning, which is relatively well developed, and local planning, which remains relatively underdeveloped (Grime and Whitelegg 1982). Local facilities should not be planned in the absence of detailed knowledge of local circumstances, but sometimes (as in the case of RAWP) national social and economic policies even contradict the aims of health care planning at any scale (Eyles et al. 1982). Perhaps most of the crucial inequalities in health are actually caused by local physical inaccessibility, so the locality is the level at which action must be taken. It is at this level that the reality of supply and demand meet and, therefore, this is the level at which detailed planning must make sense of broad national objectives.

Interdisciplinary involvement in planning: a way ahead?

Locality planning Exeter-style illustrates one way in which a number of professions and disciplines, not all of which are normally directly associated with health care, can be incorporated in the planning process. This is by no means a common practice even in Britain, however, as exemplified by the case of the Blackburn Health District, where the strategic plan was prepared, approved and implemented by health care professionals without a formal decision-making input from the consumers or even from the taxpayers who foot the bill. The strategic plan did appear conscious of the need to improve accessibility and to centralize only those facilities which medical technology rather than 'good management' demanded, although these distinctions are often debatable and unclear. As Whitelegg writes, 'in essence, the problems of inequalities in health care are inextricably bound up with the definition and delivery of a highly professionalized, remote and autocratic service which responds badly to the needs of groups and individuals outside its culturally circumscribed norms' (1982, p. 110).

This is true to the extent that existing health service planning structures in Britain, USA, Australia, USSR, and many other countries are very much professionally dominated and oriented (by doctors and professional administrators) which generally means that the 'consumer' has little say in the process. This is at least in part because the idea of a health service 'consumer' is often a misconception: in health services, a patient often comes more as a supplicant than a consumer with market rights and influences, and is often unable to understand the relative technical merits of different treatments or to make an informed choice amongst them (Stacey 1976; Ben-Sira 1976). Health services professionals, indeed, often end up thinking that the service belongs to them rather than the public or whoever happens to be paying for it. This is a pity since it reduces health care provision to an end in itself rather than a means of assisting people to achieve a better life (Torrens 1978).

Health care providers such as doctors, dentists and others can effectively regulate the demand for their services by their actions and attitudes (Heenan 1980; Phillips 1981a). In view of this and the administration dominance in health service planning, what can be achieved in the future in terms of improving flexibility and reducing inequalities in health, possibly through improving health care uptake (even if this relationship is tenuous)? Perhaps any positive action measures will be labelled as 'counterrevolutionary' by Harvey (1973) because they will not lead to the resolution of fundamental contradictions between the capitalist mode of production and the health of the people (this presumably includes current problems in socialist countries also!). Within the limitations of existing structures, though, what can be achieved?

Chapter 1 outlined the current fragmentation of academic and research efforts in health into a number of disciplines and subdisciplines, whilst the question of planning and provision of services has tended to remain professionally and administratively controlled. Even non-medical professionals, such as statisticians and some epidemiologists, have tended to be separated from the decision-making process, partly as a result of a restrictive concept of the proper place of a scientific discipline in relation to administration (E.G. Knox 1979). Therefore, we suggest that a more flexible, interdisciplinary or multidisciplinary approach to health and planning may be a way ahead.

The role of geographers in this new approach may be relatively minor. It may be relegated in some systems, perhaps, to providing information on the catchment areas or spheres of influence of hospitals and surgeries (Haynes and Bentham 1979; Phillips 1979b), or to comparing spatial configurations of service provision and disease rates (Pyle 1971; Pyle and Lauer 1975). In this case, medical and health care geography is as much an adjunct to planning as statistics and epidemiology have often been. However, if spatial scientists can be more squarely incorporated as equal partners in the planning process, evaluating existing and proposed services and suggesting alternatives, then some very positive results could occur. Geographers trained in both main 'branches' of medical geography, disease ecology and health care planning, possess potentially useful combinations of talents: epidemiological, spatial and statistical (Shannon 1980; Phillips 1981a).

It is almost impossible to conceive of any universally acceptable or applicable job specifications for health services planning. They will vary very much internationally and even within counties (for example, from urban to rural jurisdictions). Therefore, the key must be the mix of professional identities in planning, even if one of the more authoritative works on this topic does not include medical geographers, perhaps because their existence and role are poorly understood at present (E.G. Knox 1979). Instead, the following professional interactions are identified: medical statisticians and medical epidemiologists (virtually synonymous); clinical doctors and nurses; social scientists and sociologists; economists; and operational research scientists. The problem is perhaps that, in common with sociology, medical geography has been encompassing a very broad range of activities, differently interpreted by different people.

Therefore, to return to the theme introduced in Chapter 1, can there be enough common ground for interdisciplinary planning, incorporating the 'consumer', to proceed? Can the spatial scientist, in fact, contribute his considerable range of skills to this process? The role of geographers in health care planning has been quite modest to date but there are indications that it could become more substantial in the future (Giggs 1983a). This will be particularly so if researchers can be produced who are aware both of disciplinary requirements and of the very practical needs of decision-makers; and who are able to function in interdisciplinary settings (Shannon 1980; Phillips 1981a).

Certainly, some developments will occur through increasingly sophisticated analysis of data and the adoption of some of the models to optimize facility locations and accessibility outlined earlier in this book. Geographers often have skills in modelling and in the manipulation of data based on geographical subareas, and they may be employed to perform social area analyses of use in community medicine (Scott-Samuel 1977) and to analyse comprehensive, spatially gathered data banks as in the USA Health Systems Agency areas (Pyle 1979; Giggs 1983a). The very basic skill of the geographer in producing maps of health facilities and disease potential can become a most useful tool for the interdisciplinary planning team. The value of this and of other geographic skills is only just becoming apparent to many involved in this field.

L. I. H. E.
THE MARKLAND LIBRARY
STAND PARK RD., LIVERPOOL, L16 9JD

Bibliography

ABLER, R., J.S. ADAMS & P. GOULD (1971) *Spatial Organization*. Englewood Cliffs, New Jersey: Prentice-Hall.

ADAY, L.A. & R. ANDERSEN (1974) A framework for the study of access to medical care, *Health Services Research*, 9, 208–220.

AKERELE, O. (1983) Which way for traditional medicine? *World Health*, June, 3–4.

ALDERSON, M.R. (1970) Social class and the health service, *The Medical Officer*, 124, 50–52.

ALDERSON, M.R. (1976) *An Introduction to Epidemiology*. London: Macmillan.

AMBROSE, P. (1977) *Access and Spatial Inequality, Unit 23, D204, Fundamentals of Human Geography*. Milton Keynes: Open University Press.

ANDERSON, D.W. & R.M. ANDERSEN (1972) Patterns of use of health services. In H. Freeman, S. Levine and L. Reeder (eds) *Handbook of Medical Sociology*, 2nd edn. Englewood Cliffs, New Jersey: Prentice-Hall.

ANDERSON, R.A. (1968) A behavioral model of families' use of health services, *Centre for Health Administration Studies, 25, University of Chicago*.

ANDREWS, H.F. (1982) Preventive intervention for the health and well-being of urban children, *Report No. 17, The Child in the City Programme, University of Toronto*.

ARROZ, M.E. (1979) Difusão espacial da hepatite infecciosa, *Finisterra*, 14, 36–69.

ASHFORD, J.R. (1978) Regional variations in dental care in England and Wales, *British Dental Journal*, 7, 275–283.

ASHLEY, D.J.B.(1969a) A male-female difference in tumour incidence, *British Journal of Cancer*, 23, 21–25.

ASHLEY, D.J.B. (1969b) Sex differences in the incidence of tumours at various sites, *British Journal of Cancer*, 23, 26–30.

ASKEW, I.D. (1983) Modelling provision of public services in rural areas, *South West Papers in Geography No. 3, Department of Geography, University of Exeter*.

AUSTRALIAN BUREAU OF STATISTICS (1981) *Australia Year Book 1981*. Canberra: Australian Bureau of Statistics.

AVERY-JONES, F. (1976) The London hospital scene, *British Medical Journal*, 2, 1046–1049.

BABSON, J.H. (1972) *Health Care Delivery Systems: a Multinational Survey*. London: Pitman Medical.

BANERJI, D. (1979) Place of the indigenous and western systems of medicine in the health services of India, *International Journal of Health Services*, 9, 511–519.

BARLEY, S.L. (1982) Health for all by the year 2000, *Journal of the Royal College of General Practitioners*, 32, 715–716.

BARNARD, K. (1982) Retirement housing in the United Kingdom: a geographical appraisal. In A.M. Warnes (ed) *Geographical Perspectives on the Elderly*. Chichester: Wiley.

BARNETT, J.R. (1978) Race and physician location: trends in two New Zealand urban areas, *New Zealand Geographer*, 34, 2–12.

BARNETT, J.R. (1981) Reply to 'Measuring potential accessibility to general practitioners in urban areas: a methodological note', by A.E. Joseph, *New Zealand Geographer*, 37, 33–34.

BARNETT, J.R. & P. BARNETT (1977) How accessible are physicians? Some findings and implications of the changing distribution of general practitioners in three New Zealand cities, *New Zealand Medical Journal*, 85, 509–511.

BARNETT, J.R. & P. NEWTON (1977) Intra-urban disparities in the provision of primary health care: an examination of three New Zealand urban areas, *Australian and New Zealand Journal of Sociology*, 13, 60–68.

BARNETT, J.R. & I.G. SHEERIN (1977) Public policy and urban health: a review of selected programmes designed to influence the spatial distribution of physicians in New Zealand, *New Zealand Geographical Society, Proceedings of the Ninth Geographical Conference*, 10–16.

BARNETT, J.R. & I.G. SHEERIN (1978) Inefficiency and inequality: an evaluation of selected policy responses to medical maldistribution problems in New Zealand, *Community Health Studies*, 2, 65–72.

BASS, M. & W.J. COPEMAN (1975) An Ontario solution to medically underserviced areas: an evaluation of an ongoing program, *Canadian Medical Association Journal*, 113, 403–407.

BASSUK, E.L. & G. GERSON (1978) Deinstitutionalization and mental health services, *Scientific American*, 238, 46–53.

BASU, R. (1982) Use of emergency room facilities in a rural area: a spatial analysis, *Social Science and Medicine*, 16, 75–84.

BEAN, I.W. (1967) Future manpower needs in general practice, *Canadian Medical Association Journal*, 97, 1573–1577.

BEAUJEU-GARNIER, J. (1966) *Geography of Population*. London: Longman.

BEAUMONT, C.D. & L.A. PIKE (1983) General practice: a quantitative study of temporal and spatial variations in morbidity, *South West Papers in Geography No. 5, University of Exeter*.

BECK, A. (1981) *Medicine, Tradition and Development in Kenya and Tanzania, 1920-1970*. Waltham, Mass.: Crossroads Press.

BECK, R.G. (1973) Economic class and access to physician services under public medical care insurance, *International Journal of Health Services*, 3, 341–355.

BECKER, M.H. (ed) (1974) *The Health Belief Model and Personal Health Behavior*. San Francisco: Society for Public Health Education, Inc.

BENHAM, L., A. MAURIZI & W.M REDER (1968) Migration, location and remuneration of medical personnel: physicians and dentists, *Review of Economics and Statistics*, 50, 332–347.

BENNETT, D. (1979) Deinstitutionalization in two cultures, *Milbank Memorial Fund Quarterly*, 57, 516–532.

BENNETT, W.D. (1981) A location-allocation approach to health care facility location: a study of the undoctored population in Lansing, Michigan, *Social Science and Medicine*, 15D, 305–312.

BEN-SIRA, Z. (1976) The function of the professional's affective behavior in client satisfaction: a revised approach to social interaction theory, *Journal of Health and Social Behaviour*, 17, 3–11.

BERNSTEIN, J.D., F.P. HEGE & C.C. FARRAN (1979) *Rural Health Centers in the United States*. Cambridge, Mass.: Ballinger.

BIBLE, B.L. (1970) Physicians' views of medical practice in nonmetropolitan communities, *Public Health Reports*, 85, 11–17.

BIRDSALL, S.S. (1979) Planning urban service locations for the elderly. In S.M. Golant (ed) *Location and Environment of Elderly Population*. New York: Wiley, Halsted Press.

BLAXTER, M. (1976) Social class and health inequalities. In C.O. Carter and J. Peel (eds) *Equalities and Inequalities in Health*. London: Academic Press.

BOHLAND, J. & FRENCH, P. (1982) Spatial aspects of primary health care for the elderly. In A.M. Warnes (ed) *Geographical Perspectives on the Elderly*. Chichester: Wiley.

BONE, M. (1973) *Family Planning Services in England and Wales*. London: H.M.S.O.

BOOTH, A. & N. BABCHUK, (1972) Seeking health care from new resources, *Journal of Health and Social Behavior*, 13, 90–99.

BORSKY, P.N. & O. SAGEN (1959) Motivations towards health examinations, *American Journal of Public Health*, 49, 514–527.

BOSANQUET, N. (1971) Inequalities in the health service, *New Society*, 450, 809–812.

BOULDING, K.E. (1966) The concept of need for health services, *Milbank Memorial Fund Quarterly*, 44, 202–223.

BRADLEY, J.E., A.M. KIRBY & P.J. TAYLOR (1978) Distance decay and dental decay: a study of dental health among primary school children in Newcastle-upon-Tyne, *Regional Studies*, 12, 529–540.

BRADSHAW, J. (1972) A taxonomy of social need. In G. McLachlan (ed) *Problems and Progress in Medical Care*. Essays on current research, seventh series. Oxford: Oxford University Press.

BRIDGSTOCK, M. (1976) Professions and social background: the work organisation of general practitioners, *The Sociological Review*, 24, 309–329.

BROTHERSTON, J. (1976) Inequality: is it inevitable? The Galton Lecture, 1975. In C.O. Carter and J. Peel (eds) *Equalities and Inequalities in Health*. London: Academic Press.

BROWN, D.L. (1974) The redistribution of physicians and dentists in incorporated places of the upper midwest, 1950-1970, *Rural Sociology* 39, 205–223.

BROWNLEA, A.A. & C.L. WARD (1976) Health care access problems in relatively isolated communities in Northern Queensland and Darling Downs: a preliminary report, *Papers of the First Australia-New Zealand Regional Science Conference*, 174–192.

BRYANT, B.E. (1981) Issues on the distribution of health care: some lessons from Canada, *Public Health Reports*, 96, 442–447.

BUCHANAN, I.C. & I.M. RICHARDSON (1973) Time study of consultations in general practice, *Scottish Health Services Studies No. 27*, Scottish Home and Health Department.

BULMAN, J.S., N.D. RICHARDS, G.L. SLACK & A.J. WILLCOCKS (1968) *Demand and Need for Dental Care*. London: Oxford University Press.

BUTLER, J.R. (1976) How many doctors are needed in general practice? *British Medical Journal*, 1, 134–135.

BUTLER, J.R., J.M. BEVAN & R.C. TAYLOR (1973) *Family Doctors and Public Policy*. London: Routledge and Kegan Paul.

BUTLER, J.R. & R. KNIGHT (1974) *The Designated Areas Project Study of Medical Practice Areas, Final Report*. Health Services Research Unit, Centre for Research in the Social Sciences, University of Kent at Canterbury.

BUTLER, J.R. & R. KNIGHT (1976) Medical practice areas in England: some facts, and figures, *Health Trends*, 8, 8–12.

CARMICHAEL, C.L. (1983) General dental service care in the northern region, *British Dental Journal*, 154, 337–339.

CARTWRIGHT, A. (1967) *Patients and Their Doctors*. London: Routledge and Kegan Paul.

CARTWRIGHT, A. (1970) *Parents and Family Planning Services*. London: Routledge and Kegan Paul.

CARTWRIGHT, A. & R. ANDERSON (1981) *General Practice Revisited*. London: Tavistock Publications.

CARTWRIGHT, A. & M. O'BRIEN (1976) Social class variations in health care and in the nature of general practitioner consultations. In M. Stacey (ed) The sociology of the N.H.S., *Sociological Review Monographs No. 22, University of Keele*.

CARTWRIGHT, F.F. (1977) *A Social History of Medicine*. London: Longman.

CASLEY, D.J. & D.A. LURY (eds) (1981) *Data Collection in Developing Countries*. Oxford: Oxford University Press.

CHEN, M.K. (1978) A quantitative index of access to primary medical care for health planning, *Social Science and Medicine*, 12, 295–301.

CHESTER, T.E. & ICHIEN, M. (1983) Health care in Japan: its development, structure and problems, *The Three Banks Review*, 137, 17–26.

CIOCCO, A. & I. ALTMAN (1954) Medical service areas and distances travelled for physician care in western Pennsylvania, *U.S. Public Health Service, Public Health Monograph No. 19*.

CLARE, A. (1976) *Psychiatry in Dissent*. London: Tavistock.

CLEARY, P.D., D. MECHANIC & J.R. GREENLEY (1982) Sex differences in medical care utilization: an empirical investigation, *Journal of Health and Social Behavior*, 23, 106–119.

CLELAND, E.A., R.J. STIMSON & A.J. GOLDSWORTHY (1977a) Suburban health care behaviour in Adelaide, *Centre for Applied Social and Survey Research, Monograph Series 2, Flinders University, Adelaide*.

CLELAND, E.A., R.J. STIMSON & A.J. GOLDSWORTHY (1977b) Access costs of households in isolated areas in using city-based high order health care services, *ANZSERCH Proceedings*, 112–122.

COATES, B.E., R.J. JOHNSTON & P.L. KNOX (1977) *Geography and Inequality*. Oxford: Oxford University Press.

COATES, B.E. & E.M. RAWSTRON (1971) *Regional Variations in Britain: Selected Essays in Economic and Social Geography*. London: Batsford.

COCKERHAM, W.C. (1978) *Medical Sociology*. Englewood Cliffs, New Jersey: Prentice-Hall.

COMMISSION OF THE EUROPEAN COMMUNITIES (1983) Social security: a Europe-wide debate, *European File 7/83*. Brussels: Commission of the European Communities.

CONWAY, H. (1976) Emergency medical care, *British Medical Journal*, 2, 511–513.

COOPER, R., M. STEINHAUER, W.J. MILLER, R. DAVID & A. SCHATZKIN (1981) Racism, society and disease: an exploration of the social and biological mechanisms of differential mortality, *International Journal of Health Services*, 11, 389–414.

COPPO, P. (1983) Traditional psychiatry in Mali, *World Health*, June, 10–12.

COURT, S.D.M. (1976) *Fit for the Future: the Report of the Committee on Child Health Services*. Cmnd 6684. London. H.M.S.O. (The Court Report).

COX, K.R. (1979) *Location and Public Problems*. Oxford: Blackwell.

COX, K.R. & D.R. REYNOLDS (1974) Locational approaches to power and conflict. In K.R. Cox, D.R. Reynolds and S. Rokkan (eds) *Locational Approaches to Power and Conflict*. Beverley Hills, California: Sage.

CROZIER, R.C. (1968) *Traditional Medicine in Modern China*. Cambridge, Mass.: Harvard University Press.

CURTIS, S.E. (1982) Spatial analysis of surgery locations in general practice, *Social Science and Medicine*, 16, 303–313.

CUST, G. (1979) A preventive medicine viewpoint. In I. Sutherland (ed) *Health Education: Perspectives and Choices*. London: Allen and Unwin.

DADZIE, K.K.S. (1980) Economic development, *Scientific American*, 243, 58–65.

D'ARCY, A. (1976) The manufacturing and obsolescence of madness, *Social Science and Medicine*, 10, 5–13.

DARTINGTON, T. (1979) Fragmentation and integration in health care: the referral process and social brokerage, *Sociology of Health and Illness*, 1, 13–39.

DAVEY, S.C. & G.G. GILES (1979) Spatial factors in mental health care in Tasmania, *Social Science and Medicine*, 13D, 87–94.

DAVIDSSON, S.M. (1978) Understanding the growth of emergency department utilization, *Medical Care*, 16, 122–132.

DEAR, M.J., (1974) A paradigm for public facility location, *Antipode*, 6, 46–50.

DEAR, M.J. (1976) *Geographical Dimensions of the Demand for Mental Health Care*. Department of Geography, McMaster University, Hamilton, Ontario.

DEAR, M.J. (1977a) Locational factors in the demand for mental health care, *Economic Geography*, 53, 223–240.

DEAR, M.J. (1977b) Psychiatric patients and the inner city, *Annals of the Association of American Geographers*, 67, 588–594.

DEAR, M.J. (1978) Planning for mental health care: a reconsideration of public facility location theory, *International Regional Science Review*, 3, 93–111.

DEAR, M.J. & S.M. TAYLOR (1982) *Not on Our Street*. London: Pion.

DEPARTMENT OF HEALTH AND SOCIAL SECURITY (1976a) *Sharing Resources for Health in England. Report of the Resource Allocation Working Party*. London: H.M.S.O.

DEPARTMENT OF HEALTH AND SOCIAL SECURITY (1976b) *Prevention and Health: Everybody's Business*. London: H.M.S.O.

DEPARTMENT OF HEALTH AND SOCIAL SECURITY (1980) *Inequalities in Health*. Report of a Research Working Group Chaired by Sir Douglas Black. London: D.H.S.S. An edited version is available: P. Townsend and N. Davidson (1982) *Inequalities in Health*. London: Pelican.

DEPARTMENT OF HEALTH AND SOCIAL SECURITY (1981) *Care in Action*. London: H.M.S.O.

DEPARTMENT OF HEALTH AND SOCIAL SECURITY (1982) *Health and Personal Social Services Statistics for England 1982*. London: H.M.S.O.

DEPPE, H. (1977) Some remarks on the economic and political development of health care in the Federal Republic of Germany, *International Journal of Health Services*, 7, 349–357.

DESAI, M. (1974) *Marxian Economic Theory*. London: Gray-Mills.

De VISE, P. (1968) Methods and concepts of an interdisciplinary regional hospital study, *Health Services Research*, 3, 166–173.

De VISE, P. (1973) Misused and misplaced hospitals and doctors, *Commission on College Geography Resource Paper 22*. Washington: Association of American Geographers.

DEWDNEY, J.C. (1979) *A Geography of the Soviet Union*, 3rd edn. Oxford: Pergamon.

DICKEN, P. & P.E. LLOYD (1981) *Modern Western Society: A Geographical Perspective on Work, Home and Well-Being*. London: Harper and Row.

DICKSON, S. (1968) Class attitudes to dental treatment, *British Journal of Sociology*, 19, 206–211.

DIESFELD, H.J. & M.K. HECKLAU (1978) *Kenya: a Geomedical Monograph*. Berlin: Springer-Verlag.

DISEKER, R.A. & J.A. CHAPPELL (1976) Relative importance of variables in determination of practice location: a pilot study, *Social Science and Medicine*, 10, 559–563.

DOMINION BUREAU OF STATISTICS (1968) *Mental Health Statistics*. Publication No. 83-506. Ottawa: Dominion Bureau of Statistics.

DONABEDIAN, A. (1973) *Aspects of Medical Care Administration*. Cambridge, Mass.: Harvard University Press.

DOYAL, L. & I. PENNELL (1979) *The Political Economy of Health*. London: Pluto Press.

DUKES, M.N.G. (1978) Declaration of Alma Ata, *Lancet*, 2, 1256.

EARICKSON, R. (1970) The spatial behavior of hospital patients, *Department of Geography, Research Paper No. 124, University of Chicago*.

EASTON, B. (1980) *Social Policy and the Welfare State in New Zealand*. Auckland: Allen and Unwin.

EISENBERG, B.S. & J.R. CANTWELL (1976) Policies to influence the spatial distribution of physicians: a conceptual review of selected programs and empirical evidence, *Medical Care*, 14, 455–468.

ELESH, D. & P.T. SCHOLLAERT (1972) Race and urban medicine: factors affecting the distribution of physicians in Chicago, *Journal of Health and Social Behavior*, 13, 236–250.

ELLING, R.H. (1981) Political economy, cultural hegemony and mixes of traditional and modern medicine, *Social Science and Medicine*, 15A, 89–99.

ENTERLINE, P.E., V. SALTER, A.D. McDONALD & J.C. McDONALD (1973) The distribution of medical services before and after 'free' medical care — the Quebec experience, *The New England Journal of Medicine*, 289, 1174–1178.

ENTHOVEN, A. (1980) *Health Plan: The Only Practical Solution to the Soaring Cost of Medical Care*. Reading, Mass.: Addison-Wesley.

ESPING-ANDERSEN, G. (1979) Comparative social policy and political conflict in advanced welfare states: Denmark and Sweden, *International Journal of Health Services*, 9, 269–293.

EVANS, R.G., E.M.A. PARISH & F. SCULLY (1973) Medical productivity, scale effects, and demand generation, *Canadian Journal of Economics*, 6, 376–393.

EVANS, W. & M.K. CHEN (1977) The application of quantitative indices for health planning to regional health service areas in Vermont, *Social Indicators Research*, 5, 181–193.

EXETER COMMUNITY HEALTH COUNCIL (1983) *Medical Services in Rural Areas*. Exeter: Community Health Council.

EYLES, J., D.M. SMITH & K.J. WOODS (1982) Spatial resource allocation and state practice: the case of health service planning in London, *Regional Studies*, 16, 239–253.

EYLES, J. & K. WOODS (1983) *The Social Geography of Medicine and Health*. London: Croom Helm.

FAHS, I.J. & O.L. PETERSON (1965) Towns without physicians and towns with only one – a study of four states in the Upper Midwest, *American Journal of Public Health*, 58, 1200–1211.

FARIS, R.E. & H.W. DUNHAM (1939) *Mental Disorders in Urban Areas*. Chicago: University of Chicago Press.

FARMER, R.D.T. & D.L. MILLER (1977) *Lecture Notes on Community Medicine*. Oxford: Blackwell.

FARRAR, D.E. & R.R. GLAUBER (1967) Multicollinearity in regression analysis, the problem revisited, *Review of Economics and Statistics*, 49, 92–107.

FEIN, R. (1972) On achieving access and equity in health care, *Milbank Memorial Fund Quarterly*, 50, 157–190.

FELDMAN, J. (1966) *The Dissemination of Health Information*. Chicago: Aldine.

FENDALL, N.R.E. (1981) Primary health care: issues and constraints, *Third World Planning Review*, 3, 387–401.

FIEDLER, J.L. (1981) A review of the literature on access and utilization of medical care with special emphasis on rural primary care, *Social Science and Medicine*, 15C, 129–142.

FISHBEIN, M. (1967) Attitude and the prediction of behavior. In M. Fishbein (ed) *Attitude Theory and Measurement*. New York: Wiley.

FOSTER, D.P. (1976) Social class differences in sickness and in general practitioner consultations, *Health Trends*, 8, 29–32.

FOX, T.F. (1960) The 'personal doctor' and his relation to the hospital, *Lancet*, 1, 743–760.

FOX, T.F. (1962) Personal medicine, *Bulletin of the New York Academy of Medicine*, 38–527.

FOXALL, G.R. (1977) *Consumer Behaviour: a Practical Guide*, Corbridge, Northumberland: Retail and Planning Associates.

FREEMAN, H. (ed) (1983) *Mental Health and the Environment*. London: Churchill Livingstone.

FREEMAN, H., S. LEVINE & L. REEDER (eds) (1972) *Handbook of Medical Sociology*. Englewood Cliffs, New Jersey: Prentice-Hall.

FREIDSON, E. (1961) *Patients' Views of Medical Practice*. New York: Russell Sage Foundation.

FREIDSON, E. (1970) *Profession of Medicine*. New York: Dodd Mead.

FRY, J. (1971) Medical care in three societies, *International Journal of Health Services*, 1, 121–133.

FRY, J. (1973) Information for patient care in office-based practice, *Medical Care*, 11, 35–40.

FRY, J. (1979) The place of primary care. In Royal College of General Practitioners, *Trends in General Practice*. London: Royal College of General Practitioners.

GALVIN, M.E. & M. FAN (1975) The utilization of physicians' services in Los Angeles County, 1973, *Journal of Health and Social Behavior*, 16, 74–94.

GARCIA, R. (1975) *From Medical Geography to Health Geography*. Kulturgeografiskt Seminarium, Stockholm University, Stockholm.

GEORGE, P. (1959) *Questions de Géographie de la Population*. Paris: Presses Universitaires de France.

GEORGE, P. (1978) Perspectives de recherche pour la géographie des maladies, *Annales de Géographie*, 87, 641–650.

GIGGS, J.A. (1973) The distribution of schizophrenics in Nottingham, *Transactions of the Institute of British Geographers*, 59, 55–76.

GIGGS, J.A. (1979) Human health problems in urban areas. In D.T. Herbert and D.M. Smith (eds) *Social Problems and the City*. Oxford: Oxford University Press.

GIGGS, J.A. (1983a) Health. In M. Pacione (ed) *Progress in Urban Geography*. London: Croom Helm.

GIGGS, J.A. (1983b) Schizophrenia and ecological structure in Nottingham. In N.D. McGlashan and J. Blunden (eds) *Geographical Aspects of Health*. London: Academic Press.

GIGGS, J.A., D.S. EBDON & J.B. BOURKE (1980) The epidemiology of primary acute pancreatitis in the Nottingham Defined Population Area, *Transactions of the Institute of British Geographers*, N.S. 5, 229–242.

GIGGS, J.A. & P.M. MATHER (1983) Perspectives on mental health — a Nottingham example, *Report Series in Applied Geography No. 2, Department of Geography, University of Nottingham*.

GIRT, J.L. (1973) Distance to general medical practice and its effect on revealed ill-health in a rural environment, *Canadian Geographer*, 17, 154–166.

GOBER, P. & R.J. GORDON (1980) Intraurban physician location: a case study of Phoenix, *Social Science and Medicine*, 14D, 407–417.

GODLUND, S. (1961) Population, regional hospitals, transport facilities and regions: planning the location of regional hospitals in Sweden, *Lund Studies in Geography*, Series B, Human Geography No. 21. Gleerup, Lund.

GOLANT, S.M. (ed) (1979) *Location and Environment of Elderly Population.* New York: Wiley.

GOLDBERG, W.M. (1967) The present manpower situation with regard to specialists in Canada, *Canadian Medical Association Journal,* 97, 1578–1582.

GOLDSTEIN, M.S. & P.J. DONALDSON (1979) Exploiting professionalism: a case study of medical education, *Journal of Health and Social Behavior,* 20, 322–327.

GOLDSTEIN, S.G.M. (1982) In search of an optimal health care system, *British Medical Journal,* 2, 824–825, 828.

GOODMAN, A.B. & C. SIEGEL (1978) Differences in white-nonwhite community health care utilization patterns, *Journal of Evaluation and Program Planning,* 1, 51–63.

GOUGH, I. (1979) *The Political Economy of the Welfare State.* London: Macmillan.

GOULD, P. & T.R. LEINBACH (1966) An approach to the geographic assignment of hospital service, *Tijdschrift voor Economische en Sociale Geografie,* 57, 203–206.

GRAY, C. (1980) Health manpower planning: projections and pitfalls, *Canadian Medical Association Journal,* 123, 312–314.

GRIME, L.P. & J. WHITELEGG (1982) The geography of health care planning: some problems of correspondence between local and national policies, *Community Medicine,* 4, 201–208.

GROSS, P.F. (1972) Urban health disorders, spatial analysis and the economics of health facility location, *International Journal of Health Services,* 2, 63–84.

GUDGIN, G. (1975) The distribution of schizophrenics in Nottingham: a comment, *Transactions of the Institute of British Geographers,* 60, 148–149.

GUPTILL, S.C. (1975) The spatial availability of physicians, *Proceedings of the Association of American Geographers,* 7, 80–84.

GUZICK, D.S. (1978) The demand for general practitioner and internist services, *Health Services Research,* 351–368.

HADLEY, J. (1979) Alternative methods of evaluating health manpower distribution, *Medical Care,* 17, 1054–1060.

HALIBURTON, KAWARTHA & PINE RIDGE DISTRICT HEALTH COUNCIL (1977) *Inventory of Physicians by Specialty and Location.* Peterborough, Ontario: Haliburton, Kawartha and Pine Ridge District Health Council.

HALLAS, J. (1976) *CHCs in Action.* London: Nuffield Provincial Hospitals Trust.

HAMNETT, C. (1979) Area-based explanations: a critical appraisal. In D.T. Herbert and D.M. Smith (eds) *Social Problems and the City.* Oxford: Oxford University Press.

HANKOFF, L., C.J. RABNER & C.G. HENRY (1971) Comparison of the satellite clinic and the hospital-based clinic, *Archives of General Psychiatry,* 24, 474–478.

HARDIMAN, M. & J. MIDGLEY (1982) Social planning and access to the social services in developing countries: the case of Sierra Leone, *Third World Planning Review,* 4, 74–86.

HART, J.T. (1971) The inverse care law, *Lancet,* 1, 405–412.

HARVEY, D. (1973) *Social Justice and the City.* London: Edward Arnold.

HARVEY, D. (1983) *Limits to Capital.* Oxford: Blackwell.

HAYNES, R.M. (1983) The geographical distribution of mortality by cause in Chile, *Social Science and Medicine*, 17, 355–364.

HAYNES, R.M. & C.G. BENTHAM (1979) *Community Hospitals and Rural Accessibility*. Farnborough: Saxon House.

HAYNES, R.M. & C.G. BENTHAM (1982) The effects of accessibility on general practitioner consultations, out-patient attendances and in-patient admissions in Norfolk, England, *Social Science and Medicine*, 16, 561–569.

HEALTH RESEARCH GROUP (ed) (1982) Contemporary perspectives on health and health care, *Occasional Paper No. 2, Department of Geography, Queen Mary College, London*.

HEENAN, L.D.B. (1980) Health service planning and projected population change: some observations for New Zealand, *Social Science and Medicine*, 14D, 241–249.

HEIBY, J.R. (1982) Primary health care: some lessons from Nicaragua, *World Health Forum*, 3, 27–29.

HERBERT, D.T. (1972) *Urban Geography: a Social Perspective*. Newton Abbot: David and Charles.

HERBERT, D.T. (1979) Geographical perspectives and urban problems. In D.T. Herbert and D.M. Smith (eds) *Social Problems and the City*. Oxford: Oxford University Press.

HERBERT, D.T. & S.M. PEACE (1980) The elderly in an urban environment: a study of Swansea. In D.T. Herbert and R.J. Johnston (eds) *Geography and the Urban Environment*. Chichester: Wiley.

HERBERT, D.T. & D.M. SMITH (eds) (1979) *Social Problems and the City*. Oxford: Oxford University Press.

HERBERT, D.T. & C.J. THOMAS (1982) *Urban Geography: a First Approach*. Chichester: Wiley.

HETZEL, B.S. (1980) *Health and Australian Society*. Harmondsworth: Penguin.

HEUMANN, L. & D. BOLDY (1982) *Housing for the Elderly*. London: Croom Helm.

HINES, R. (1972) The health status of black Americans: changing perspectives. In E. Jaco (ed) *Patients, Physicians and Illness*. New York: Free Press.

HITZHUSEN, F. & T. NAPIER (1978) A rural public service policy. In D.L. Rogers and L.R. Whiting (eds) *Rural Policy Research Alternatives*. Ames, Iowa: Iowa State University Press.

HOLLINGSWORTH, J.R. (1981) Inequality in levels of health in England and Wales, 1891–1971, *Journal of Health and Social Behavior*, 22, 268–283.

HONIGSBAUM, F. (1972) Quality in general practice, *Journal of the Royal College of General Practitioners*, 22, 429–449.

HOPKINS, E.G., A.M. PYE, M. SOLOMON & S. SOLOMON (1968) The relation of patients' age, sex and distance from surgery to the demand on the family doctor, *Journal of the Royal College of General Practitioners*, 16, 368–378.

HORNER, A. & A. TAYLOR (1979) Grasping the nettle: locational strategies for Irish hospitals, *Administration*, 27, 348–370.

HORTON, F.E. & D.R. REYNOLDS (1969) An investigation of individual action spaces, *Proceedings of the Association of American Geographers*, 1, 70–75.

HORTON, F.E. & D.R. REYNOLDS (1971) Effects of urban spatial structure on individual behavior, *Economic Geography*, 47, 36–46.

HOWE, G.M. (1963) *National Atlas of Disease Mortality in the United Kingdom*. London: Nelson.

HOWE, G.M. (1972) *Man, Environment and Disease in Britain*. Newton Abbot: David and Charles.

HOWE, G.M. (ed) (1977) *A World Geography of Human Diseases*. London: Academic Press.

HOWE, G.M. (1980) Medical geography. In E.M. Brown (ed) *Geography Yesterday and Tomorrow*. Oxford: Oxford University Press.

HOWE, G.M. (1982) Does it matter where I go? *Proceedings of the Royal Society of Edinburgh*, 82B, 75–96.

HOWE, G.M. & D.R. PHILLIPS (1983) Medical geography in the United Kingdom, 1945–1982. In N.D. McGlashan and J. Blunden (eds) *Geographical Aspects of Health*. London: Academic Press.

HUFF, D.L. (1960) A topographic model of consumer space preferences, *Papers and Proceedings of the Regional Science Association*, 6, 159–174.

HUNT, J. (1957) The renaissance of general practice, *British Medical Journal*, 1, 1075–1082. Reprinted in *Journal of the Royal College of General Practitioners, Supplement No. 4*, 1972.

ILLICH, I. (1976) *Limits to Medicine*. Harmondsworth: Penguin.

INGRAM, D. (1971) The concept of accessibility: a search for an operational form, *Regional Studies*, 5, 101–107.

INGRAM, D.R., D.R. CLARKE & R.A. MURDIE (1978) Distance and the decision to visit an emergency department, *Social Science and Medicine*, 12, 55–62.

INSTITUTE OF HEALTH SERVICES ADMINISTRATORS (1982) *The Hospitals and Health Services Year Book 1982*. London: Institute of Health Services Administrators.

IRVING, H. (1975) A geographer looks at personality, *Area*, 7, 207–212.

ISARD, W. (1960) *Methods of Regional Analysis: an Introduction*. Cambridge, Mass.: MIT Press.

JAGGI, O.P. (1976) *Allopathy, Homoeopathy, Ayurveda, Unani and Nature Cure*. Delhi: Orient Paperbacks.

JEFFERY, R. (1982) Policies towards indigenous healers in independent India, *Social Science and Medicine*, 16, 1835–1841.

JOHNSTON, R.J. (1978) *Multivariate Statistical Analysis in Geography*. London: Longman.

JONES, K. & A. KIRBY (1982) Provision and well-being: an agenda for public resources research, *Environment and Planning A*, 14, 297–310.

JOROFF, S. & V. NAVARRO (1971) Medical manpower: a multivariate analysis of the distribution of physicians in urban United States, *Medical Care*, 9, 428–438.

JOSEPH, A.E. (1979) The referral system as a modifier of distance decay effects in the utilization of mental health care services, *Canadian Geographer*, 23, 159–169.

JOSEPH, A.E. (1981) Measuring potential accessibility to general practitioners in urban areas: a methodical note, *New Zealand Geographer*, 37, 32–33.

JOSEPH, A.E. (1982) On the interpretation of the coefficient of localization, *Professional Geographer*, 34, 443–446.

JOSEPH, A.E. & P.R. BANTOCK (1982) Measuring potential physical accessibility to general practitioners in rural areas: a method and case study, *Social Science and Medicine*, 16, 85–90.

JOSEPH, A.E. & P.R. BANTOCK (1983) Rural accessibility of general practitioners: a Canadian perspective, *Paper presented at the I.B.G. Annual Conference, Edinburgh, Scotland*.

JOSEPH, A.E. & J.L. BOECKH (1981) Locational variation in mental health care utilization dependent upon diagnosis: a Canadian example, *Social Science and Medicine*, 15D, 395–404.

JOSEPH, A.E. & A. POYNER (1981) The utilization of three public services in a rural Ontario township: an empirical evaluation of a conceptual framework, *Centre for Resources Development Publication No. 107, University of Guelph, Ontario*.

JOSEPH, A.E. & A. POYNER (1982) Interpreting patterns of public service utilization in rural areas, *Economic Geography*, 58, 262–273.

JOSEPH, A.E. & B. SMIT (1981) Implications of exurban residential development: a review, *Canadian Journal of Regional Science*, 4, 207–224.

KALISZER, M. & M. KIDD (1981) Some factors affecting attendance at ante-natal clinics, *Social Science and Medicine*, 15D, 421–424.

KAMERLING, D.S. (1976) Regional inequality in the availability of health care in the Soviet Union in 1970, *Proceedings of the Association of American Geographers*, 8, 125–129.

KANE, R.L. (1969) Determination of health care priorities and expectations among rural consumers, *Health Services Research*, 4, 142–151.

KENNEDY, V.C. (1979) Rural access to a regular source of medical care, *Journal of Health*, 4, 199–203.

KERR, R.B. (1967) Future manpower needs in specialty practice, *Canadian Medical Association Journal*, 97, 1583–1586.

KIKHELA, N., G. BIBEAU & E. CORIN (1981) Africa's two medical systems: options for planners, *World Health Forum*, 2, 96–99.

KINSTON, W. (1983) Pluralism in the organization of health services, *Social Science and Medicine*, 17, 299–313.

KIRK, R.F.H. & R.M. SPEARS (1979) Development of rural health services: problems illustrated by a nonprofit, privately-based approach, *Medical Care*, 17, 175–182.

KLEINMAN, A., P. KUNSTADTER, E.R. ALEXANDER & J.L. GALE (eds) (1975) Medicine in Chinese cultures, *National Institute of Health Publication NIH 75–653*. Washington: U.S. Department of Health, Education and Welfare.

KNOX, E.G. (ed) (1979) *Epidemiology in Health Care Planning*. Oxford: Oxford University Press.

KNOX, P.L. (1975) *Social Well-Being: A Spatial Perspective*. Oxford: Oxford University Press.

KNOX, P.L. (1978) The intraurban ecology of primary medical care: patterns of accessibility and their policy implications, *Environment and Planning A*, 10, 415–435.

KNOX, P.L. (1979a) Medical deprivation, area deprivation and public policy, *Social Science and Medicine*, 13D, 111–121.

KNOX P.L. (1979b) The accessibility of primary care to urban patients: a geographical analysis, *Journal of the Royal College of General Practitioners*, 29, 160–168.

KNOX, P.L. (1982a) *Urban Social Geography*. London: Longman.

KNOX, P.L. (1982b) The geography of medical care delivery: an historical perspective, *Geoforum*, 12, 245–250.

KNOX, P.L. & J. BOHLAND (1983) The locational dynamics of medical care settings in American Cities 1860–1940, *Paper presented at the I.B.G. Annual Conference, Edinburgh, Scotland*.

KNOX, P.L., J. BOHLAND & N.L. SHUMSKY (1983) The urban transition and the evolution of the medical care delivery system in America, *Social Science and Medicine*, 17, 37–43.

KNOX, P.L. & M. PACIONE (1980) Locational behaviour, place preferences and the inverse care law in the distribution of primary medical care, *Geoforum*, 11, 43–55.

KOHN, R. & K.L. WHITE (eds) (1976) *Health Care: an International Study*. Oxford: Oxford University Press.

KOOS, E. (1954) *The Health of Regionsville*. New York: Columbia University Press.

KROEGER, A. (1982) Los indígenas sudamericanos ante una alternativa: servicios de salud tradicionales o modernos, en las zonas rurales de Ecuador, *Boletín de la Oficina Sanitaria Panamericana*, 93, 300–315.

KROEGER, A. (1983) Anthropological and socio-medical health care in developing countries. *Social Science and Medicine*, 17, 147–161.

KRUPINSKI, J. & A. STOLLER (eds) (1971) *The Health of a Metropolis*. Melbourne: Heinemann Educational Books.

LANKFORD, P. (1971) The changing location of physicians, *Antipode*, 3, 68–72.

LAW, C.J. & A.M. WARNES (1976) The changing geography of the elderly in England and Wales, *Transactions of the Institute of British Geographers*, N.S. 1, 453–471.

LAWSON, R.J. (1980) Patients' attitudes to doctors, *Journal of the Royal College of General Practitioners*, 30, 137–138.

LAWTON, M.P. (1970) Planning environments for older people, *Journal of the American Institute of Planners*, 36, 124–129.

LEARMONTH, A.T.A. (1972) Medicine in medical geography. In N.D. McGlashan (ed) *Medical Geography: Techniques and Field Studies*. London: Methuen.

LEARMONTH, A.T.A. (1978) *Patterns of Disease and Hunger*. Newton Abbot: David and Charles.

LEARMONTH, A.T.A. (1982) Prevention is better than cure, *Geographical Magazine*, 54, 638–641.

LEARMONTH, A.T.A. & R. AKHTAR (1979) India's malaria resurgence 1965–1978 *Geography*, 64, 221–223.

LEE, R.P.L. (1975) Health service systems in Hong Kong: professional stratification in a modernizing society, *Inquiry*, 12, suppl. 51.

LEE, R.P.L. (1981) Chinese and western medical care in China's rural commune: a case study, *Social Science and Medicine*, 15A, 137–148.

LEE, R.P.L. (1982) Comparative studies of health care systems, *Social Science and Medicine*, 16, 629–642.

LEFROY, R.B. & J. PAGE (1972) Assessing the needs of elderly people, *Medical Journal of Australia*, 2, 1071–1075.

LLOYD, P.E. & P. DICKEN (1968) The data bank in regional studies of industry, *Town Planning Review*, 38, 304–316.

LLOYD, R.E. (1977) Consumer behavior after migration: a reassessment process, *Economic Geography*, 53, 14–27.

LOMAS, H.D. & J.D. BERMAN (1983) Diagnosing for administrative purposes: some ethical problems, *Social Science and Medicine*, 17, 241–244.

LONDON HEALTH PLANNING CONSORTIUM (1980) *Towards a Balance: a Framework for Acute Hospital Services in London*. London: H.M.S.O.

LUCAS, A. (1980) Changing medical models in China: organizational options or obstacles? *The China Quarterly*, 83, 461–489.

McBROOM, W.H. (1970) Illness, illness behavior and socioeconomic status, *Journal of Health and Social Behavior*, 11, 319–326.

McCORMICK, J.S. (1976) The 'personal doctor' 1975, *Journal of the Royal College of General Practitioners*, 26, 750–753.

McEWIN, R. (1981) Health services in Australia, *World Hospitals*, 17, 10–15.

McGLASHAN, N.D. (ed) (1972) *Medical Geography: Techniques and Field Studies*. London: Methuen.

McGLASHAN, N.D. & J.R. BLUNDEN (eds) (1983) *Geographical Aspects of Health*. London: Academic Press.

McKELVEY, D.J. (1979) Transportation issues and problems of the rural elderly. In S.M. Golant (ed) *Location and Environment of Elderly Population*. New York: Wiley.

McKINLAY, J.B. (1970) The new late comers for antenatal care, *British Journal of Preventive and Social Medicine*, 24, 52–57.

McKINLAY, J.B. (1971) The concept 'patient career' as a heuristic device for making medical sociology relevant to medical students, *Social Science and Medicine*, 5, 441–460.

McKINLAY, J.B. (1972) Some approaches and problems in the study of the use of services — an overview, *Journal of Health and Social Behavior*, 13, 115–152.

MACLEAN, U. & R.H. BANNERMAN (1982) Utilization of indigenous healers in national health delivery systems, *Social Science and Medicine*, 16, 1815–1816.

MacMAHON, B. & T.F. PUCH (1970) *Epidemiology: Principles and Methods*. Boston: Little Brown.

MADDOX, G.L. (1971) Muddling through: planning for health care in England, *Medical Care*, 9, 439–448.

MAGNUSSON, G. (1980) The role of proximity in the use of hospital emergency departments, *Sociology of Health and Illness*, 2, 202–214.

MAIR, L. (1965) *An Introduction to Social Anthropology*. Oxford: Clarendon Press.

MANDEL, E. (1973) *An Introduction to Marxist Economic Theory*, 2nd edn. New York: Pathfinder Press.

MARDEN, P.G. (1966) A demographic and ecological analysis of the distribution of physicians in metropolitan America, 1960, *American Journal of Sociology*, 72, 290–300.

MASSAM, B.H. (1974) Political geography and the provision of services, *Progress in Geography*, 6, 179–210.

MASSAM, B.H. (1975) *Location and Space in Social Administration*. London: Edward Arnold.

MASSAM, B.H. (1980) *Spatial Search*. Oxford: Pergamon Press.

MAXWELL, R. (1980) *International Comparisons of Health Needs and Services*. London: King's Fund Centre.

MAY, J.M. (1950) Medical geography: its methods and objectives, *Geographical Review*, 40, 9–41.

MAY, J.M. (1958) *The Ecology of Human Disease*. New York: MD Publications.

MAYER, J.D. (1982) Relations between two traditions of medical geography, *Progress in Human Geography*, 6, 216–230.

MAYNARD, A. (1975) *Health Care in the European Community*. London: Croom Helm.

MAYNARD, A. (1983) Privatizing the National Health Service, *Lloyds Bank Review*, 148, 28–41.

MECHANIC, D. (1962) The concept of illness behavior, *Journal of Chronic Diseases*, 15, 189–194.

MECHANIC, D. (1968) *Medical Sociology: a Selective View*. New York: Free Press.

MECHANIC, D. (1970) Problems in the future organization of medical practices, *Law and Contemporary Problems*, 35, 233–248.

MECHANIC, D. (1972) *Public Expectations and Health Care*. New York: Wiley Interscience.

MERCER, J. (1979) Locational consequences of housing policies for the low-income elderly: a case study. In S.M. Golant (ed) *Location and Environment of Elderly Population*. New York: Wiley.

MERGET, A.E. (1980) Equity in the distribution of municipal services. In H.J. Bryce (ed) *Revitalizing Cities*. Lexington, Mass.: Lexington Books.

MILLAS, A.J. (1980) Planning for the elderly within the context of a neighbourhood, *Ekistics*, 283, 264–273.

MILLER, A.E. (1977) The changing structure of the medical profession in urban and suburban settings, *Social Science and Medicine*, 11, 233–243.

MILLER, D.H. (1974) *Community Mental Health: A Study of Services and Clients*. Lexington, Mass.: D.C. Heath.

MISHAN, E.J. (1964) *Welfare Economics: Five Introductory Essays*. New York: Random House.

MISHAN, E.J. (1969) *Welfare Economics: an Assessment*. Amsterdam: North Holland Publishing Co.

MONTEIRO, L. (1973) Expense is no object....: Income and physician services reconsidered, *Journal of Health and Social Behavior*, 14, 99–115.

MORAN, W. & S.J. NASON (1981) Spatio-temporal localization of New Zealand dairying, *Australian Geographical Studies*, 19, 47–66.

MORRELL, D.C., H.G. GAGE & N.A. ROBINSON (1970) Patterns of demand in general practice, *Journal of the Royal College of General Practitioners*, 19, 331–342.

MORRILL, R.L. & R.J. EARICKSON (1968a) Variation in the character and use of hospital services, *Health Services Research*, 3, 224–238.

MORRILL, R.L. & R.J. EARICKSON (1968b) Hospital variation and patient travel distance, *Inquiry*, 5, 26–34.

MORRILL, R.L., R.J. EARICKSON & P. REES (1970) Factors influencing distances travelled to hospitals, *Economic Geography*, 46, 161–171.

MORRILL, R.L. & M. KELLEY (1970) The simulation of hospital use and the estimation of location efficiency, *Geographical Analysis*, 2, 283–300.

MOSCOVICE, I., C.W. SCHWARTZ & S.M. SHORTELL (1979) Referral patterns of family physicians in an underserviced rural area, *Journal of Family Practice*, 9, 677–682.

MOSELEY, M.J. (1979) *Accessibility: The Rural Challenge*. London: Methuen.

MOTT, F.D. & M.I. ROEMER (1948) *Rural Health and Medical Care*. New York: McGraw-Hill.

MOUNTJOY, A.B. (ed) (1978) *The Third World: Problems and Perspectives*. London: Macmillan.

MURRAY, G.D. & A. CLIFF (1977) A stochastic model for measles epidemics in a multi-region setting, *Transactions of the Institute of Medical Geographers*, N.S. 2, 158–174.

NATH, S.K. (1973) *A Perspective of Welfare Economics*. London: Macmillan.

NATHANSON, C.A. (1975) Illness and the feminine role: a theoretical review, *Social Science and Medicine*, 9, 57–62.

NAVARRO, V. (1976) *Medicine under Capitalism*. New York: Prodist.

NEUMANN, A.K. & T. LAURO (1982) Ethnomedicine and biomedicine linking, *Social Science and Medicine*, 16, 1817–1824.

NORMAN, P. (1975) *Managerialism: Review of Recent Work*. Conference Paper No. 14. London: Centre for Environmental Studies.

NORTHCOTT, H.C. (1980) Convergence or divergence: the rural-urban distribution of physicians and dentists in census divisions and incorporated cities, towns and villages in Alberta, Canada, 1956–1976, *Social Science and Medicine*, 14D, 17–22.

OFFICE OF HEALTH ECONOMICS (1974) *The NHS Reorganization*. (No. 48 in series on current health problems.) London: Office of Health Economics.

OFFICE OF HEALTH ECONOMICS (1977) *The Reorganized NHS*. (No. 58 in series on current health problems.) London: Office of Health Economics.

OFFICE OF POPULATION CENSUSES AND SURVEYS (1970) *Classification of Occupations*. London: H.M.S.O.

OHLSEN. S., A.L. RICKETTS & J.T. CURRAN (1980) Problems in the evaluation of a community health centre in Tasmania, *Social Science and Medicine*, 14D, 267–269.

O'MULLANE, D.M. & M.E. ROBINSON (1977) The distribution of dentists and the uptake of dental treatment by schoolchildren in England, *Community Dentistry and Oral Epidemiology*, 5, 156–159.

ORUBULOYE, I.O. & O.Y. OYENEYE (1982) Primary health care in developing countries: the case of Nigeria, Sri Lanka and Tanzania, *Social Science and Medicine*, 16, 675–686.

PAHL, R.E. (1970) *Whose City?* London: Longman.

PARKER, R.C. & T.G. TUXILL (1967) The attitudes of physicians toward small-community practice, *Journal of Medical Education*, 42, 327–344.

PARKIN, D. (1979) Distance as an influence on demand in general practice, *Epidemiology and Community Health*, 33, 96–99.

PARMELEE, D.E., G. HENDERSON & M.S. COHEN (1982) Medicine under socialism: some observations on Yugoslavia and China, *Social Science and Medicine*, 16, 1389–1396.

PARRY, W.H. (1979) *Communicable Diseases*. London: Hodder and Stoughton.

PEACE, S.M. (1982) The activity patterns of elderly people in Swansea, South Wales, and South-East England, In A.M. Warnes (ed) *Geographical Perspectives on the Elderly*. Chichester: Wiley.

PERKIN, R.L. (1967) Medical manpower in general practice, *Canadian Medical Association Journal*, 97, 1569–1572.

PHILLIPS, D.R. (1979a) Spatial variations in attendance at general practitioner services, *Social Science and Medicine*, 13D, 169–181.

PHILLIPS, D.R. (1979b) Public attitudes to general practitioner services: a reflection of an inverse care law in intraurban primary medical care? *Environment and Planning A*, 11, 815–824.

PHILLIPS, D.R. (1980) Spatial patterns of surgery attendance: some implications for the planning of primary health care, *Journal of the Royal College of General Practitioners*, 30, 688–695.

PHILLIPS, D.R. (1981a) *Contemporary Issues in the Geography of Health Care*. Norwich: Geo Books.

PHILLIPS D.R. (1981b) The planning of social service provision in the new towns of Hong Kong, *Planning and Administration*, 8, 8–23.

PHILLIPS, D.R. (1983a) Service provision in new towns: the Hong Kong example, *Geografisch Tijdschrift*, 17, 326–337.

PHILLIPS, D.R. (1983b) Medical services in new towns where 'mixed' traditional and modern systems exist: Hong Kong's example. In Singapore Professional Centre, *High-rise, High-density Living*. Singapore: Professional Centre.

PHILLIPS, D.R. & M. COURT (1982) "Grassrooting" — locality planning in health services, *The Health Services*, Oct. 29, 14–15.

PHILLIPS, D.R. & A.M. WILLIAMS (1984) *Rural Britain: a Social Geography*. Oxford: Blackwell.

PHILLIPS, D.R. & A.G.O. YEH (1983) China experiments with modernization: the Shenzhen Special Economic Zone, *Geography*, 68, 289–300.

PICHERAL, H. (1976) *Espace et santé. Géographie médicale du Midi de la France*. Montpellier: Paysan du Midi.

PICHERAL, H. (1982) Géographie médicale, géographie des maladies, géographie de la santé, *L'Espace Géographique*, 11, 161–175.

PILLSBURY, B.L.K. (1982) Policy and evaluation perspectives on traditional health practitioners in national health care systems, *Social Science and Medicine*, 16, 1825–1834.

PINCH, S. (1979) Territorial justice and the city: a case study of the social services for the elderly in Greater London. In D.T. Herbert and D.M. Smith (eds) *Social Problems and the City*. Oxford: Oxford University Press.

PIRIE, G.H. (1983) On spatial justice, *Environment and Planning A*, 15, 465–473.

POLITICS OF HEALTH GROUP (1982) *'Going private'. The Case Against Private Medicine*. London: Politics of Health Group and Fightback.

POOLE, M.A. & P.N. O'FARRELL (1971) The assumptions of the linear regression model, *Transactions of the Institute of British Geographers*, 52, 145–158.

PORAPAKKHAM, Y. (1982) Thailand case studies on sex differences in utilization of health resources, *Publication No. 5, Institute for Population and Social Research, Mahidol University, Thailand.*

PYLE, G.F. (1971) Heart disease, cancer and stroke in Chicago, *Department of Geography Research Paper No. 124, University of Chicago.*

PYLE, G.F. (1979) *Applied Medical Geography*. New York: Wiley.

PYLE, G.F. & B.M. LAUER (1975) Comparing spatial configurations: hospital service areas and disease rates, *Economic Geography*, 51, 50–68.

RAMESH, A. & B. HYMA (1981) Traditional Indian medicine in practice in an Indian metropolitan city, *Social Science and Medicine*, 15D, 69–81.

REILLY, B.J., J.S. LEGGE & M.S. REILLY (1980) A rural health perspective: principles for rural health policy, *Inquiry*, 17, 120–127.

REIN, M. (1969) Social class and the health service, *New Society*, 20 Nov., 807–810.

REINHARDT, U.E. (1982) Table manners at the health care feast, In D. Yaggy (ed) *Financing Health Care: Competition or Regulation*. Cambridge, Mass: Ballinger.

REYNOLDS, M. (1978) No news is bad news: patients' views about communication in hospital, *British Medical Journal*, 1, 1673–1676.

RICKARD, J.E. (1976) Per capita expenditure of English Area Health Authorities, *British Medical Journal*, 1, 299–300.

RIMLINGER, G.V. & H.B. STEELE (1963) An economic interpretation of the spatial distribution of physicians in the United States, *Southern Economic Journal*, 30, 1–12.

RITCHIE, J., A. JACOBY & M. BONE (1981) *Access to Primary Health Care*. London: H.M.S.O.

ROBERTS, D.F. (1976) Sex differences in disease and mortality. In C.O. Carter and J. Peel (eds) *Equalities and Inequalities in Health*. London: Academic Press.

ROBERTSON, I.M.L. (1978) Planning the location of recreation centres in an urban area: a case study of Glasgow, *Regional Studies*, 12, 419–427.

ROBINSON, W.S. (1950) Ecological correlations and the behavior of individuals, *American Sociological Review*, 15, 351–357.

ROEMER, M. (1948) Historical development of the current crisis in rural medicine in the U.S.. In S.R. Kagan (ed) *Victor Robinson Memorial Volume: Essays on History of Medicine*. New York: Froben Press.

ROEMER, M.I. (1977) *Systems of Health Care*. New York: Springer.

ROEMER, M.I. & R.J. ROEMER (1981) *Health Care Systems and Comparative Manpower Policies*. New York: Marcel Dekker.

ROGERS, G.F., R.W. TRAVERS & G.P. MALANSON (1980) An insular geography approach to equilibrium numbers of physician specialties across urban centres, *Social Science and Medicine*, 14D, 45–54.

ROGHMANN, K.J. & T.R. ZASTOWNY (1979) Proximity as a factor in the selection of health care providers: emergency room visits compared to obstetric admissions and abortions, *Social Science and Medicine*, 13D, 61–69.

ROOS, N.P., M. GAUMONT & J.M. HORNE (1976) The impact of the physician surplus on the distribution of physicians across Canada, *Canadian Public Policy*, 2, 169–191.

ROSE, H.M. (1971) *The Black Ghetto: a Spatial and Behavioral Perspective*. New York: McGraw-Hill.

ROSEN, G. (1968) *Madness in Society*. New York: Harper and Row.

ROSENFIELD, S. (1980) Sex differences in depression: do women always have higher rates? *Journal of Health and Social Behavior*, 21, 33–42.

ROSENSTOCK, I.M. (1960) What research in motivation suggests for public health, *American Journal of Public Health*, 50, 295–302.

ROSENSTOCK, I.M. (1966) Why people use health services, *Health Services Research, I, The Milbank Memorial Fund Quarterly*, 44, 94–127.

ROSENSTOCK, I.M., M. DERRYBERRY & B.K. CARRIGER (1959) Why people fail to seek poliomyelitis vaccination, *Public Health Reports*, 74, 98–103.

ROTHMAN, D.J. (1980) *Conscience and Convenience: The Asylum and its Alternatives in Progressive America*. Boston: Little Brown.

ROYAL COLLEGE OF GENERAL PRACTITIONERS (1972) How many patients? *Journal of the Royal College of General Practitioners*, 22, 491–493.

ROYAL COLLEGE OF GENERAL PRACTITIONERS (1979) *Trends in General Practice, 1979*. London: Royal College of General Practitioners.

ROYAL COLLEGE OF GENERAL PRACTITIONERS (1982) Health for all by the year 2000, *Journal of the Royal College of General Practitioners*, 32, 715–717.

RYAN, M. (1978) *The Organization of Soviet Medical Care*. Oxford: Blackwell.

SALKEVER, D.S. (1975) Economic class and differential access to health care, *International Journal of Health Services*, 5, 373–395.

SALMOND, G.C. (1973) Social needs for medical services: the inverse care law in New Zealand, *New Zealand Medical Journal*, 80, 396–403.

SALMOND, G.C. (1974) Medical manpower planning, *New Zealand Medical Journal*, 80, 459–462.

SANKAR, D.V.S. & J. MINTUS (1978) An analysis of telephone referrals for a mental health association chapter, *Social Science and Medicine*, 12, 63–66.

SANSOM, C.D., J. WAKEFIELD & R. YULE (1972) Cervical cytology in the Manchester area: changing patterns of response. In J. Wakefield (ed) *Seek Wisely to Prevent*. London: H.M.S.O.

SCARPA, A. (1981) Pre-scientific medicines: their extent and value, *Social Science and Medicine*, 15A, 317–326.

SCARROTT, D.M. (1978) Changes in the regional distribution of general dental service manpower, *British Dental Journal*, June 6, 359–363.

SCHMITT, R.R. (1979) Transportation and the urban elderly: local problems, ameliorative strategies, and national policies. In S.M. Golant (ed) *Location and Environment of Elderly Population*. New York: Wiley.

SCHNEIDER, J.B. (1967) Measuring the locational efficiency of the urban hospital, *Health Services Research*, 2, 154–159.

SCHULTZ, R.R. (1975) A space potential analysis of physician location, *Proceedings of the Association of American Geographers*, 7, 203–208.

SCOTT-SAMUEL, A. (1977) Social area analysis in community medicine, *British Journal of Preventive and Social Medicine*, 31, 199–204.

SCULL, A.T. (1978) *Decarceration: Community Treatment of the Deviant — a Radical View*. Englewood Cliffs, New Jersey: Prentice-Hall.

SÈNE, P. (1983) The TBAs of Senegal, *World Health*, June, 22–25.

SHANNON, G.W. (1977) Space, time and illness behavior, *Social Science and Medicine*, 11, 683–689.

SHANNON, G.W. (1980) The utility of medical geography research, *Social Science and Medicine*, 14D, 1–2.

SHANNON, G.W., R.L. BASHSHUR & C.A. METZNER (1969) The concept of distance as a factor in accessibility and utilization of health care, *Medical Care Review*, 26, 143–161.

SHANNON, G.W. & G.E.A. DEVER (1974) *Health Care Delivery: Spatial Perspectives*. New York: McGraw-Hill.

SHANNON, G.W., J. LOVETT & R.L. BASHSHUR (1979) Travel for primary care: expectation and performance in a rural setting, *Journal of Community Health*, 5, 113–125.

SHANNON G.W., J.L. SKINNER & R.L. BASHSHUR (1973) Time and distance: the journey for medical care, *International Journal of Health Services*, 3, 237–244.

SHANNON, G.W. & C.W. SPURLOCK (1976) Urban ecological containers, environmental risk cells, and the use of medical services, *Economic Geography*, 52, 171–180.

SHANNON, G.W., C.W. SPURLOCK, S.T. GLADIN & J.L. SKINNER (1975) A method for evaluating the geographic accessibility of health services, *Professional Geographer*, 27, 30–36.

SHAW, S. (1975) The role of the nurse in assessing the health of elderly people. In G. McLachlan (ed) *Probes for Health*. Nuffield Provincial Hospitals Trust. London: Oxford University Press.

SHORTELL, S.M. (1972) *A Model of Physician Referral Behavior: a Test of Exchange Theory in Medical Practice*. Chicago: Centre for Health Administration Studies, University of Chicago.

SIDEL, V.W. & R. SIDEL (1979) Health care services as part of China's revolution and development. In N.G. Maxwell (ed) *China's Road to Development*, 2nd edn. Oxford: Pergamon.

SMIT, B. & A.E. JOSEPH (1982) Trade-off analysis of preferences for public services. *Environment and Behavior*, 14, 238–258.

SMITH, C.J. (1976) Distance and the location of community mental health facilities: a divergent viewpoint, *Economic Geography*, 52, 181–191.

SMITH, D.M. (1974) Who gets what *where* and how: a welfare focus for human geography, *Geography*, 59, 289–297.

SMITH, D.M. (1977) *Human Geography: a Welfare Approach*. London: Edward Arnold.

SMITH, D.M. (1979) *Where the Grass is Greener: Living in an Unequal World*. Harmondsworth: Penguin.

SOHLER, K.B. & J.D. THOMPSON (1970) Jarvis' law and the planning of mental health services, *Public Health Reports*, 85, 503–510.

SORRE, M. (1933) Complexes pathogènes et géographie médicale, *Annales de Géographie*, 235, 1–18.

SORRE, M. (1943) *Les Fondements Biologiques de la Géographie Humaine*. Paris: Librairie Armand Colin.

SORRE, M. (1947) *Les Fondements de la Géographie Humaine*. Paris: Librairie Armand Colin.

SORRE, M. (1966) Géographie de l'état sanitaire et des maladies, *Encyclopédie de la Pléiade. Géographie Générale*, 1089–1106.

SPAULDING, W.B. & W.O. SPITZER (1972) Implications of medical manpower trends in Ontario, 1961–1971, *Ontario Medical Review*, 39, 527–533.

STACEY, M. (1976) The health services consumer: a sociological misconception. In M. Stacey (ed) The sociology of the NHS, *Sociological Review Monograph No. 22, University of Keele*.

STACEY, M. (1977) People who are affected by the inverse law of care, *Health and Social Services Journal*, June 3, 898–902.

STARR, P. (1977) Medicine, economy and society in nineteenth-century America, *Journal of Social History*, 10, 588–607.

STEINMETZ, N. & J.R. HOEY (1978) Hospital emergency room utilization in Montreal before and after Medicare, *Medical Care*, 16, 133–139.

STIMSON, R.J. (1980) Spatial aspects of epidemiological phenomena and the utilization of health care services in Australia: a review of methodological problems and empirical analyses, *Environment and Planning A*, 12, 881–907.

STIMSON, R.J. (1981)The provision and use of general practitioner services in Adelaide, Australia: application of tools of locational analysis and theories of provider and user spatial behaviour, *Social Science and Medicine*, 15D, 27–44.

STIMSON, R.J. (1983) Research design and methodological problems in the geography of health. In N.D. McGlasha and J. Blunden, (eds) *Geographical Aspects of Health*. London: Academic Press.

STIMSON, R.J. & E.A.CLELAND (1975) Household health care behavior in suburban areas of Adelaide: a baseline study. In J.S. Dodge and S.R. West (eds) *Epidemiology and Primary Medical Care*. Dunedin: University of Otago.

STOCK, R. (1981) Traditional healers in rural Hausaland, *Geojournal*, 5, 363–368.

SUCHMAN, E.A. (1964) Sociomedical variations among ethnic groups, *American Journal of Sociology*, 70, 319–331.

SUCHMAN, E.A. (1965a) Social patterns of illness and medical care, *Journal of Health and Human Behavior*, 6, 2–16.

SUCHMAN, E.A. (1965b) Stages of illness and medical care, *Journal of Health and Human Behavior*, 6, 114–128.

SUCHMAN, E.A. (1966) Health orientation and medical care, *American Journal of Public Health*, 56, 97–105.

SUMNER, G. (1971) Trends in the location of primary medical care in Britain: some social implications, *Antipode*, 3, 46–53.

SUTHERLAND, I. (ed) (1979) *Health Education: Perspectives and Choices.* London: Allen and Unwin.

TARLO, K.G. (1980) Equity and local autonomy in service provision, *Social Science and Medicine*, 14D, 233–235.

TAYLOR, C.E. (1976) The place of indigenous medical practitioners in the modernization of health services. In C. Leslie, (ed) *Asian Medical Systems: a Comparative Study.* Berkeley: University of California Press.

TAYLOR, D.G., L.A. ADAY & R. ANDERSEN (1975) A social indicator of access to medical care, *Journal of Health and Social Behavior*, 16, 38–49.

TAYLOR, P.J. (1977a) *Quantitative Methods in Geography: An Introduction to Spatial Analysis.* London: Houghton-Mifflin.

TAYLOR, P.J. (1977b) *Distance Decay in Spatial Interactions.* CATMOG 2. Norwich: Geo Abstracts Ltd.

TAYLOR, P.J. & C.L. CARMICHAEL (1980) Dental health and the application of geographical methodology, *Community Dentistry and Oral Epidemiology*, 8, 117–122.

THOMAS, C.J. (1974) The effects of social class and car ownership on intra-urban shopping behaviour in Greater Swansea, *Cambria*, 1, 98–126.

THOMAS, C.J. (1976) Sociospatial variation and the use of services. In D.T. Herbert and R.J. Johnston (eds) *Social Areas in Cities, Volume 2.* Chichester: Wiley.

THORNHILL NEIGHBOURHOOD PROJECT (1978) *Health Care in Thornhill: a Case of Inner City Deprivation.* London: Thornhill Neighbourhood Project.

THOUEZ, J.P. (1978) *Espace Régional et Santé: la Géographie Hospitalière des Cantons de l'Est (Québec).* Sherbrooke, Quebec: Editions Naamen.

TISCHLER, G.L., J. HENISZ, J.K. MYERS & V. GARRISON (1972) Catchmenting and the use of mental health services, *Archives of General Psychiatry*, 27 389–392.

TITMUSS, R. (1968) *Commitment to Welfare.* London: Allen and Unwin.

TODARO, M.P. (1977) *Economic Development in the Third World.* London: Longman.

TOPLEY, M. (1975) Chinese and western medicine in Hong Kong: some social and cultural determinants of variation, interaction and change. In A. Kleinman, P. Kunstadter, E.R. Alexander and J.L. Gale (eds) Medicine in Chinese cultures, *National Institute of Health Publication NIH 75-653.* Washington: U.S. Department of Health, Education and Welfare.

TORNQVIST, G., S. NORDBECK & P. GOULD (1971) Multiple location analysis, *Lund Studies in Geography*, Series C, No. 12. Gleerup, Lund.

TORRENS, P.R. (1978) *The American Health Care System: Issues and Problems*. St Louis: C.V. Mosby Co.

TORRENS, P.R. & D.G. YEDVAB (1970) Variations among emergency room populations: a comparison of four hospitals in New York City, *Medical Care*, 8, 60–75.

TREAS, J. (1977) Family support for the aged: some social and demographic considerations, *Gerontologist*, 17, 486–491.

TRIPP, P. (1981) A comparative analysis of health care costs in three selected countries: the United States, the United Kingdom and Australia, *Social Science and Medicine*, 15C, 19–30.

VEEDER, N. (1975) Health services utilization models for human services planning, *Journal of the American Institute of Planners*, 41, 101–109.

VERBRUGGE, L.M. (1979) Female illness rates and illness behavior: testing hypotheses about sex differences in health, *Women and Health*, 4, 61–79.

VERBRUGGE, L.M. & R.P. STEINER (1981) Physician treatment of men and women patients. Sex bias or appropriate care? *Medical Care*, 19, 609–632.

VERHASSELT, Y. (1975) *Maps of Cancer Distribution*. Geografisch Instituut. Brussels: Vrije Universiteit.

VILLAR, J. (1982) La mujer en la salud y el desarollo. Algunos problemas de salud de la mujer en el Tercer Mundo, *Boletín de la Oficina Sanitaria Panamericana*, 93, 327–340.

VUORI, H. (1982a) Primary health care in industrialized countries, *Journal of the Royal College of General Practitioners*, 32, 729–735.

VUORI, H. (1982b) The World Health Organization and traditional medicine, *Community Medicine*, 4, 129–137.

WACHS, M. & T.G. KUMAGAI (1973) Physical accessibility as a social indicator, *Socio Economic Planning Sciences*, 7, 437–456.

WADDINGTON, I. (1977) The relationship between social class and the use of health services in Britain, *Journal of Advanced Nursing*, 2, 609–619.

WALMSLEY, D.J. (1978) The influence of distance on hospital usage in rural New South Wales, *Australian Journal of Social Issues*, 13, 72–81.

WALMSLEY, D.J. & I.R. McPHAIL (1976) The geography of hospital care in New South Wales, *Research Series in Applied Geography No. 44*, Department of Geography, The University of New England, Armindale.

WARNES, A.M. (1981) Towards a geographical contribution to gerontology, *Progress in Human Geography*, 5, 317–341.

WARNES, A.M. (ed) (1982) *Geographical Perspectives on the Elderly*. Chichester: Wiley.

WEBB, H.L. (1982) Socialism and health in France, *Social Policy and Administration*, 16, 241–252.

WHILE, A.E. (1978) The vital role of the cottage-community hospital, *Journal of the Royal College of General Practitioners*, 28, 485–491.

WHITE, A.N. (1979) Accessibility and public facility location, *Economic Geography*, 55, 18–35.

WHITELEGG, J. (1982) *Inequalities in Health Care: Problems of Access and Provision*. Retford: Straw Barnes Press.

WIBBERLEY, G.P. (1978) Mobility and the countryside. In R. Cresswell (ed) *Rural Transport and Country Planning*. London: Leonard Hill.

WILENSKI, P. (1976) Integration of the traditional Chinese practitioner into the medical system. In *The Delivery of Health Services in the People's Republic of China*. Ottawa: International Development Research Centre.

WILLIAMS, J.I. & E.J. LUTTERBACH (1976) The changing boundaries of psychiatry in Canada, *Social Science and Medicine*, 10, 15–22.

WILLIAMS, R., M. BLOOR, G. HOROBIN & R. TAYLOR (1980) Remoteness and disadvantage: findings from a survey of access to health services in the Western Isles, *Scottish Journal of Sociology*, 4, 105–124.

WILSON, T. & D.J. WILSON (1982) *The Political Economy of the Welfare State*. London: Allen and Unwin.

WISEMAN, R.F. (1979) Regional patterns of elderly concentration and migration. In S.M. Golant (ed) *Location and Environment of Elderly Population*. New York: Wiley.

WORLD BANK (1981) *World Development Report 1981*. New York: Oxford University Press.

WORLD HEALTH ORGANIZATION (1978a) *Primary Health Care: Alma Ata 1978*. Geneva: World Health Organization.

WORLD HEALTH ORGANIZATION (1978b) *The Promotion and Development of Traditional Medicine*, WHO Technical Report Series 622, 9. Geneva: World Health Organization.

WORLD HEALTH ORGANIZATION (1980) *Sixth Report on the World Health Situation 1973–1977*. Geneva: World Health Organization.

WORLD HEALTH ORGANIZATION (1981) *Global Strategy for Health for all by the Year 2000*. Geneva: World Health Organization.

YEATES, M.H. (1974) *An Introduction to Quantitative Analysis in Human Geography*. New York: McGraw-Hill.

YONG, Y. & A. COTTERELL (1977) *Chinese Civilization*. London: Weidenfeld and Nicolson.

YUDKIN, J.S. (1978) Changing patterns of resource allocation in a London Health District, *British Medical Journal*, 2, 1212–1215.

Index of Names

Index of Subjects